Smart Agriculture

Volume 17

The book series Smart Agriculture presents progress of smart agricultural technologies, which includes, but not limited to, specialty crop harvest robotics, UAV technologies for row crops, innovative IoT applications in plant factories, and big data for optimizing production process. It includes both theoretical study and practical applications, with emphasis on systematic studies. AI technologies in agricultural productions will be emphasized, consisting of innovative algorithms and new application domains. Additionally, new crops are emerging, such as hemp in U.S., and covered as well. This book series would cover regions worldwide, such as U.S., Canada, China, Japan, Korea, and Brazil.

The book series Smart Agriculture aims to provide an academic platform for interdisciplinary researchers to provide their state-of-the-art technologies related to smart agriculture. Researchers of different academic backgrounds are encouraged to contribute to the book, such as agriculture engineers, breeders, horticulturist, agronomist, and plant pathologists. The series would target a very broad audience – all having a professional related to agriculture production. It also could be used as textbooks for graduate students.

Yinyan Shi • Xiaochan Wang • Man Chen
Lei Wang • Xuekai Huang • Zhao Zhang
Editors

Technological Progress on Variable Rate Fertilization

Springer

Editors
Yinyan Shi
Nanjing Agricultural University
Nanjing, Jiangsu, China

Man Chen
Nanjing Institute of Agriculture
Mechanization, Ministry of Agriculture and
Rural Affairs
Nanjing, Jiangsu, China

Xuekai Huang
Nanjing Agricultural University
Nanjing, Jiangsu, China

Xiaochan Wang
Nanjing Agricultural University
Nanjing, Jiangsu, China

Lei Wang
Nanjing Agricultural University
Nanjing, Jiangsu, China

Zhao Zhang
Research Center for Smart Agriculture
China Agricultural University
Beijing, China

ISSN 2731-3476 ISSN 2731-3484 (electronic)
Smart Agriculture
ISBN 978-981-95-4329-8 ISBN 978-981-95-4330-4 (eBook)
https://doi.org/10.1007/978-981-95-4330-4

This Springer imprint is published by the registered company Springer Nature Singapore Pte Ltd.
The registered company address is: 152 Beach Road, #21-01/04 Gateway East, Singapore 189721, Singapore

Preface

To meet the increasing high-quality living standards, enough food is required, and precision and smart agricultural technology plays an important role in achieving this. In both precision and smart agriculture technology, variable-rate fertilization (VRF) is a key technique defined as tailoring fertilizer application to the specific needs of different areas within a field. To achieve satisfactory VRF performance, a vast number of studies have been conducted over the past decades, including how to simulate the VFR process to improve performance and how to design the key components of a centrifugal variable-rate fertilizer spreader for better performance. Though many specific studies have been conducted, there lacks a book systematically reviewing the related technological progress, which drove the development of the current book.

This book consists a total of eight chapters: Chaps. 1 and 2 cover the progress and development on biological information of crop phenotype research applied to real-time variable-rate fertilization and precision variable-rate fertilization for rice-wheat cropping using outer-grooved wheel mechanism based on a multilayer perceptron model, respectively. Chapter 3 discusses the design and modeling of the centrifugal variable-rate fertilizer spreader. Chapter 4 addresses the theoretical study related to the mechanical properties of fertilizer particles and crop straw under discrete element modeling. Chapters 5 and 6 present the performance analysis and optimization of centrifugal variable-rate fertilizer spreaders and performance evaluation and lag compensation of actuators in variable-rate fertilizer spreaders, respectively. Chapter 7 describes the fertilizer discharge flow rate detection system, while the final chapter reports the field evaluation and efficiency analysis of centrifugal variable-rate fertilization in precision agriculture. The total of eight chapters are individually separated and are closely related to address multiple key problems related to VRF technology from both a theoretical and a practical perspective.

The editors of this book, Drs. Yinyan Shi, Xiaochan Wang, Man Chen, Lei Wang, Xuekai Huang, and Zhao Zhang, are experts in variable-rate fertilization and have worked in this field for decades, which guarantees the quality of this book *Technology Progress on Variable-Rate Fertilization: Theoretical Research and Practical Applications*.

Nanjing, Jiangsu, China	Yinyan Shi
Nanjing, Jiangsu, China	Xiaochan Wang
Nanjing, Jiangsu, China	Man Chen
Nanjing, Jiangsu, China	Lei Wang
Nanjing, Jiangsu, China	Xuekai Huang
Beijing, China	Zhao Zhang

Contents

Chapter 1
Progress and Development on Biological Information of Crop Phenotype Research Applied to Real-Time Variable-Rate Fertilization

Yinyan Shi, Xiaochan Wang, S. Hu, J. Wang, Z. Zhao, and J. Zheng

Abstract Variable-rate fertilization, as a crucial link in the implementation of Precision Agriculture (PA), can remarkably complete reasonable and effective fertilizer application and nutrient management based on local conditions, so as to obtain the maximum grain output with the minimum fertilizer input and achieve the optimal input-output production ratio. As the main operation form of variable-rate fertilization, real-time variable-rate control technology adjusts the fertilizer application online to variable-rate input as needed according to the current growth status and nutrient information of crops, and its basic core lies in the accuracy and stability of the obtained crop phenotype biological information. Due to the affinity relationship between biological information of crop phenotypic and crop real-time fertilizer demand, it has been widely used in the scientific research on key technologies of real-time variable fertilization in recent years. Mainly from the three significant aspects of inversion model, detection technology and comprehensive application for crop phenotype status, this study analyzed the establishment process and species characteristics of the inversion models between crop phenotypic biological information and nutritional status, summarized the real-time monitoring principle and key technologies of crop phenotypic biological parameters, and emphatically

Y. Shi · X. Wang (✉)
College of Engineering, Nanjing Agricultural University, Nanjing, China

Jiangsu Province Engineering Lab for Modern Facility Agriculture Technology & Equipment, Nanjing, China
e-mail: shiyinyan@njau.edu.cn; wangxiaochan@njau.edu.cn

S. Hu · J. Wang · J. Zheng
College of Engineering, Nanjing Agricultural University, Nanjing, China
e-mail: 2023112024@stu.njau.edu.cn; wangjihao@stu.njau.edu.cn; 2021212006@stu.njau.edu.cn

Z. Zhao
College of Information and Electrical Engineering, China Agricultural University, Beijing, China
e-mail: zhaozhangcau@cau.edu.cn

Y. Shi et al. (eds.), *Technological Progress on Variable Rate Fertilization*, Smart Agriculture 17, https://doi.org/10.1007/978-981-95-4330-4_1

expounded the application situation and existing problems of real-time variable-rate fertilization technology based on crop phenotypic biological parameters. And then the future emphasis research direction and development proposal of the variable-rate fertilization technology based on crop growth information were discussed combined with the needs of sustainable development of modern agriculture in China. It provides a theoretical reference for the construction of a scientific management technology system of fertilizer application reducing and effect increasing, and for the development of relevant technical equipments in line with China's unique national conditions.

Keywords Biological information · Crop phenotype · Precision agriculture · Real-time and online · Variable-rate fertilization

1.1 Introduction

Scientific, rational and effective fertilizer application plays a very positive role in tapping crop-yield potential, promoting grain production and ensuring food security, and is the only way and inevitable demand for realizing the sustainable development of modern agriculture [1, 2]. Variable-rate fertilization, as the representative pivotal technology foundation of precision agriculture, has opened up a new effective way to meet the above-mentioned green planting concept of modern agricultural with on-demand input and balanced fertilization [3]. Variable rate technology (VRT), by changing the large-scale extensive fertilization management mode of traditional agriculture, is supported by the "3S" technology (Remote Sensing RS, Geographical Information System GIS, and Global Positioning System GPS), according to the real-time crop-growth status and nutrient profit-loss to purposeful on-demand variable-rate input of fertilizer, thus balancing the soil fertility, reducing the regional differences of biochemical parameters in the field, improving fertilizer utilization rate and crop productivity capacity, obtaining the highest crop yield and best economic benefits, and promoting the mechanization, information, automation and intelligent sustainable development of modern agricultural production in China [4, 5].

Since the concept of "Precision Agriculture" (PA) was advocated and implemented by American agricultural producers in the 1980s, variable-rate fertilization technology has developed rapidly in intelligent agricultural equipment innovation field in foreign developed countries [6, 7], and its core control technology is mainly manifested two forms of based on prescription information and based on real-time parameters (as shown in Fig. 1.1) [8, 9]. The variable-rate fertilization control method based on prescription information, because of the limitations of complicated process, difficult operation and high economic cost, cannot apply variable fertilizer scientifically and reasonably according to the real-time nutrient demand of crop growth cycle [10]; however, the variable-rate fertilization control technology based on real-time parameter, on the basis of avoiding the above-mentioned tedious process, can well solve the problems of uneven growth and imbalance soil nutrient

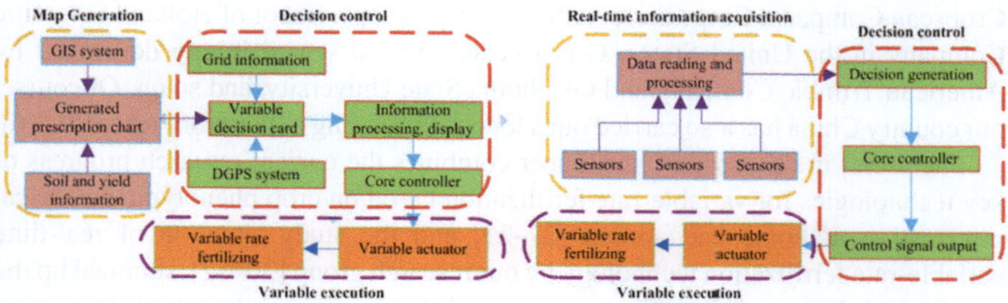

a. Based on prescription information b. Based on real-time parameters

Fig. 1.1 Technical flowchart of variable rate fertilization. (**a**) Based on prescription information. (**b**) Based on real-time parameters

in crop growth process. Relying on the advanced modern real-time sensor equipment, soil nutrient parameters and crop growth information are online monitored to generate target fertilizer application in real time [11], and scientifically guide the operation equipment to achieve on-demand variable fertilization.

As for real-time variable-rate fertilization technology, the critical point lies in the detection accuracy, representativeness and stability of crop phenotypic biological information [12]. Studies have shown that there is a certain logical relationship between the real-time fertilizer requirement of crops and the phenotypic nutritional status of their corresponding stages, and the real-time growth of crops is closely related to the biological elements of leaves, while the content of crop biological elements has the significant correlation with the canopy spectral reflectance information [13, 14]. The spectral reflectivity and spectral parameters of different wavelengths have the different quantitative relationships with the leaf nutrient elements (nitrogen, phosphorus, potassium, etc.), chlorophyll index (SPAD), leaf area index (LAI), ratio vegetation index (RVI), normalized difference vegetation index (NDVI) and other indexes can well predict the biological parameters of crop phenotype [15–17], reflect the real-time growth information of crops, and guide scientific and reasonable topdressing operations. Therefore, the fertilizer application required for crops can be estimated by the effective models of statistical analysis based on its phenotypic biological parameters [18, 19].

In recent years, with the continuous development of computer science and remote sensing (RS) technology, new ideas and methods for real-time non-destructive nutrient detection in crop growth process have been provided [20, 21]. Thus, the portable high-throughput phenotypic monitoring platforms combining agricultural remote sensing technology and crop growth bio-index have emerged at the historic moment. By carrying different types of real-time sensors, it is possible to quickly, accurately and non-destructively obtain the multi-source RS data symbolized plant biological information in a short time, determine the optimal amount of topdressing application [22, 23], and achieve the rational and on-demand regulation management of variable-rate fertilization for crops. At present, many mature products have been widely used in real-time variable-rate spreading operation equipment [24, 25], mainly including MSR-16 radiometer produced by American

Cropscan Company, Crop Circle ACS-470 crop canopy sensor of Holland Scientific Company in the United States, Greenseeker spectral sensor jointly developed by American Trimble Company and Oklahoma State University, and so on. Of course, our country China has also carried out a lot corresponding exploratory research [26].

Therefore, in this context, this paper combines the current research progress of key technologies for variable-rate fertilization based on crop phenotypic biological information at home and abroad [27–29] and the study situation of real-time variable-rate fertilization technology by our research group [30–32], summed up the species and characteristics of inversion models between crop phenotypic biological information and nutritional status, analyzed and compared the real-time monitoring principles and key technologies of crop phenotypic biological parameters, evaluated and discussed the application status and development trend of real-time variable-rate fertilization technology based on crop phenotypic biological parameters, the existing problems were pointed out and feasible suggestions were put forward to provide theoretical references for future development of real-time variable-rate fertilization technology based on crop growth information applicable to China [33].

1.2 Transmission Mechanism Between Crop Biological Information and Variable-Rate Fertilization Technology

Since its inception in the early twentieth century, remote sensing technology as a new type of advanced practical detection technology, because of its unique advantages of rapid, non-destructive, multi-platform, multi-resolution (time and space), and large-area access to ground and terrain information [34], has been widely used in many fields through long-term rapid and stable development, among which agricultural production is one of the most important areas of remote sensing technology application. Agricultural remote sensing technology plays a special role in agricultural resource survey, biological yield estimation, agricultural disaster prediction and assessment, and others. Especially in recent years, in view of the dispersion of crop distribution in China and regional complexity, the advantages of agricultural remote sensing technology are brought into play, timely, objective and accurate access to crop area, growth, yield and other information appears to be particularly important in China [33].

At present, agricultural remote sensing technology has been widely used to measure land coverage and crop biomass, etc., especially for the extraction of key biological and physicochemical parameters of crop phenotype, its theoretical basis is the target detection of spectral characteristics difference of electromagnetic waves based on crop phenotype [35, 36]. Research found that crops under the action of electromagnetic waves the spectral reflectance, absorption, transmission and emission characteristics responded in different wave bands are different, especially in the visible-near-infrared spectrum bands, the reflectivity is mainly affected by phenotypic pigments, cell structure, water content other biological parameters. There have strong absorption characteristics in the visible red light wave bands and strong reflection characteristics in the near-infrared wave bands, which is often used for monitoring crop growth, production yield and quality, pests and diseases, etc [37, 38]. Generally, the spectral reflectivity is the ratio of the reflected energy of an object to the incident energy at a particular wavelength interval. When the solar spectrum with a wavelength of λ is projected into the canopy leaves of the crops in parallel, Lambertian Reflection is produced [39], and the reflection model of crop phenotype conforms to the Lambert Cosine Law:

$$\frac{d\phi_\lambda}{dA \cdot d\Omega} = L_\lambda \cdot cos\theta \tag{1.1}$$

where φ_λ is the radiant flux, that is the radiant energy passing through a section in unit time; L_λ is the radiation luminance of crop canopy reflection; A is the projection area of crop canopy; Ω is the solid angle; θ is the reflection angle.

According to the Lambertian law above, the total radiated power per unit area of crop canopy radiation to the hemispherical space can be calculated, that is, the radiation exitance Me, and the relationship between reflection radiation luminance L_λ of crop canopy can be derived as follows:

$$\begin{aligned} M_e &= \frac{d\phi_\lambda}{dA} = L_\lambda \cdot cos\theta d\Omega = \int_0^{2\pi}\int_0^{\frac{\pi}{2}} L_\lambda \cdot cos\theta \cdot sin\theta d\theta d\Omega \\ &= \int_0^{2\pi}\int_0^{\frac{\pi}{2}} L_\lambda \cdot \frac{\sin 2\theta}{4} \cdot d(2\theta) d\Omega = \pi L_\lambda \end{aligned} \tag{1.2}$$

If the radiant flux of solar spectrum with a wavelength of λ parallel to unit area of crop canopy is E_λ, then the spectral reflectivity ρ_λ of the crop canopy with respect to the solar spectrum is

$$\rho_\lambda = \frac{M_e}{E_\lambda} = \frac{\pi L_\lambda}{E_\lambda} \tag{1.3}$$

Real-time variable-rate fertilization technology is just exactly based on these radiation transmission mechanism between the biological information of crop phenotype and crop nutrient growth and production yield quality, depends on the determination model of a nutrient element inversed by the remote sensing technology, such as the nitrogen/chlorophyll concentration, nitrogen/chlorophyll accumulation, leaf area index, biomass, density and other nutrient parameters [40]. And with the expert decision system built into the variable-rate fertilization equipment for support, scientific and rational nutrient regulation and fertilizer management are completed, effectively solve the problems of regional differences of biochemical parameters of field crops, improve fertilizer utilization efficiency and crop production capacity, and lead precision agriculture modernization and sustainable development.

1.3 Inversion Method Between Biological Information of Crop Phenotype and Nutritional Status

With the continuous development of agricultural remote sensing technology, more and more applications have been applied in the monitoring of crop phenotypic biological information and the diagnosis of plant growth nutrients in large-scale modern agricultural production, including the growth detection and yield prediction of wheat, corn, rice, cotton and other staple crops [41, 42]. Based on the monitoring of crop canopy biology information, according to the internal relation between crop biocharacteristics and canopy spectral reflectance characteristics [43, 44], through the spectral data mining: such as statistical discriminant, correlation analysis, regression model, etc. to establish the corresponding algorithm relationship, and then the inversion of crop growth information (plant nutrient status) can be acquired [45–47], such as: crop Nitrogen Nutrition Index (NNI), Ratio Vegetation Index (RVI), Leaf Nitrogen Accumulation (LNA), etc.

At this stage, the commonly used inversion methods are mainly divided into two categories [33, 48]: first one is the empirical statistical method based on the relationship between crop phenotypic biological information parameters and spectral reflectivity or other spectral characteristic parameters in sensitivity bands, which is simple and easy to use, but the established empirical relationship model excessively relies on some specific conditions such as instrument, time, variety and environment, so the applicability is poor [49]. Second one is the radiation transmission model inversion method based on the remote sensing monitoring physical process, commonly used numerical optimization algorithm, partial least squares method, principal component analysis algorithm, support vector machine training set algorithm, artificial neural network algorithm, etc. are applied to simplify the derivation process and improve inversion efficiency, which has a strong universality [50, 51]. Using this method for the research on high-throughput biological information monitoring and nutrient status discrimination of crop phenotypes have covered multiple scales from crop population to tissue and organ, and the research results are rich and diverse [52].

Crop canopy coverage, as a phenotypic parameter that quantitatively describing the dynamic differences in growth and development between genotypes, is an important indicator to characterize and evaluate crop nutrient status, such as crop nitrogen nutrient status, early vitality, senescence state, surface biomass and others, and has an important application in monitoring crop growth [53]. Currently, there are many indexes that are closely related to crop canopy coverage studied at home and abroad, among which normalized difference vegetation index (NDVI) and leaf area index (LAI) are the most mature ones [54, 55]. NDVI, which is monitored in real time, can not only provide an objective method for indirectly assessing the variation characteristics of wheat under stress state and the seasonal nitrogen demand application, used to estimate crop nitrogen content, surface nitrogen distribution and crop nitrogen utilization efficiency, and can utilize the close correlation between NDVI and crop physiological characteristics to explain the diversity impact factors of crops, such as the change regularity of moisture content, nitrogen content during crop growth stages [56–58]. Li Weiguo et al. [59] discussed the relationship between NDVI value based on TM (Thematic Mapper) image and crop leaf area index LAI, directly used the inversed LAI as the growth grading standard to monitor the growth of winter wheat in Xinghua City, Jiangsu Province, and to assess the index of crops nitrogen content and nitrogen demand application. And then combined with variable-rate fertilization technology to guide scientific and rational variable application of nitrogen fertilizer on the demand, improving the quality level of fertilizer utilization efficiency, and the economic benefit is significant.

Crop phenotype nitrogen nutrition and nitrogen demand application are the core content for monitoring and diagnosing crop growth, and the important mainstay for making management decisions of field production, providing the necessary technical basis for achieving the modernization and sustainable development of precision agriculture [60, 61]. Through experimental studies, Shibayama et al. found that the linear combination of nitrogen content per unit area of the crop leaves with the reflectivity in spectral bands R620 and R760 had good regression relationship, as well as with the reflectivity in spectral bands R400, R620 and R880, and not affected by the type of the variety [62]. Liu liangyun et al. used the characteristics of the reflection peak at 1690 nm to well invert the carbon-nitrogen ratio of wheat canopy at each growth stage, the segmented remote sensing inversion model of leaf and stem-sheath at the rise stage and milk stage were established, and the accuracy of estimating the nitrogen content in fresh leaves using spectral analysis method was improved to more than 85% [63]. Cho et al. proposed a linear extrapolation method to calculate the red edge position information of crop canopy reflectance spectrum, and the prediction models for predicting nitrogen concentration information of wheat plants, maize leaves and mixed pasture plants were established [64]. Chen et al. study showed that the dual-peak canopy nitrogen index (DCNI) eliminates the interference of leaf area changes on the diagnosis of nitrogen concentration, thus achieving the purpose of accurately detecting plant nitrogen concentration [65]. Chen Pengfei et al. realized the remote sensing estimation of nitrogen nutrition index (NNI) of winter wheat and maize, which could well determine the nitrogen nutrition status of winter wheat and maize [66].

With the deepening of studies on the monitoring of crop phenotype nitrogen nutrition by domestic and foreign scholars, more and more attention has been paid to the heterogeneity of vertical distribution of canopy nitrogen content, due to the nitrogen content at different height of crop canopy plays an important role in guiding scientific variable-rate fertilization and field management [67, 68]. Wang jihua et al. used partial least squares algorithm (PLS) to invert the nitrogen concentration of leaves at different levels based on the canopy spectra vertically observed of winter wheat [69]. Wang zhijie estimated the nitrogen density of different leaf layers by using the difference vegetation index (DVI) of winter wheat combined with the interception of luminous energy at each level of the canopy [70]. Zhao Chunjiang et al. constructed the inversion index of chlorophyll concentration in the upper, middle and lower leaf layers of winter wheat based on the observational multi-angle canopy reflectance spectra, and achieved better accuracy [71]. Ma Jifeng et al. based on field experiment data, analyzed the correlation between the spectral and fluorescence characteristics of leaves in different vertical heights and the nitrogen concentration in leaves, which provided a new idea for exploring the remote sensing method of vertical distribution of crop canopy nitrogen [72].

Using the field working machines, observation towers and artificial hand-held sensors as carriers, such as radiometer MSR-16, nitrogen sensor Yara N-sensor, crop canopy sensor Crop Circle ACS-470, spectrum sensor Greenseekers, non-imaging objects spectrometer ASD Fieldspc FR2500, etc., according to the acquired crop phenotypic biological information and the inversed nutrient evaluation models, rational nitrogen on-demand application and nitrogen regulation management can be achieved [73–75]. Kitchen et al. applied the Green Channel Normalized Differential Vegetation Index value GNDVI obtained by Crop Circle to monitor the crops nitrogen deficiency and then guided the variable-rate fertilization. The results showed that the method could reduce the nitrogen application and improve the crop yield [76]. Peng et al. proposed a mode of real-time nitrogen management (RTNM), which is based on the measured crop SPAD value to conduct the rapid diagnosis of nitrogen nutrition status. And the nitrogen fertilizer application strategy was determined by comparing and analyzing the real-time measurement results with the set threshold [77]. Scarf et al. at Missouri State University judged the nitrogen nutrition status of maize in the unfertilized area according to the difference in corn greenness value between the saturated fertilization area and the fertilization control area, and carried out the variable-rate fertilization management [78]. Lukina et al. used canopy NDVI values to invert the nitrogen accumulation of winter wheat plants and estimate the potential yield for the variable-rate fertilization management [79]. Chen Liping et al., using the measured field object spectral data, improved the nitrogen application optimization algorithm based on the spectral data proposed by Lukina et al., making it suitable for variable-rate fertilization in Beijing area and expanding the universality of the algorithm [80]. Zhao Chunjiang et al. established a variety of variable-rate fertilization algorithms for winter wheat based on the Optimized Soil-adjusted Vegetation Index OSAVI and chlorophyll SPAD values of winter wheat, and compared with the traditional algorithm based on the soil fertility for comparative study, the results showed that the economic benefits and ecological benefits of variable-rate fertilization based on the spectral data of winter wheat are superior to other fertilization algorithms [81].

1.4 Detection Technologies of Crop Phenotype Biological Information

Detection accuracy of crop phenotypic biological information plays a crucial and vital role in the effectiveness of inversion model for crop nitrogen nutrient and the feasibility of implementation of variable-rate fertilization, it is must be on the basis of information detected accurately to complete fast, high-throughput and non-destructive testing process, which has put forward higher requirements for the monitoring technology of phenotypic information [82, 83]. At present, the representative information monitoring techniques applied to crop phenotypes are divided into imaging technology and non-imaging sensing technology [84, 85]. Imaging technologies mainly include: visible light or RGB (Red-Green-Blue) imaging technology, spectral imaging technology, hyperspectral imaging technology, infrared thermal imaging technology, etc., non-imaging sensing technologies mainly include: radar sensing technology, active non-imaging sensing technology (depth sensing technology, ultrasonic technology), etc., and the commonly used sensors include color digital camera, multi-spectral camera, hyperspectral camera, depth camera, thermal imager or photosynthetic fluorescence sensor, radar or laser sensor, and so on.

1.4.1 Visible-Light Imaging Technique

Visible-light imaging technology, also known as RGB imaging technology, mostly composed of Charge Coupled Device (CCD) sensor or metal oxide semiconductor, is the most commonly used imaging technology for high-throughput phenotypic information monitoring of field crop, with the advantages low cost, intuitive and convenient access to crop phenotypic information, which can be used to monitor crop emergence rate, flowering condition, canopy coverage and lodging situation in agricultural production process [84–86]. Digital camera is based on this visible-light imaging principle to collect the information of the object in visible light wavebands, obtaining the color information and texture information of the object, and through two-dimensional or three-dimensional imaging to acquire the morphological parameters related to crop growth and yield, including leaf area index, vegetation coverage, vegetation index and plant and flower count, etc [87–89]. The crop 3 Dimension structure can be reconstructed by multi-frame images acquired by the stereo vision system, and the stereo vision system is composed of one or more color digital cameras, enabling the color digital camera to measure more phenotypic parameters [90, 91]. On account of the limitations of a large number of plant leaves overlapping and background influence on the image, the media occlusion and changes in ambient light will lead to under-exposure or over-exposure of the image, so RGB imaging technology is limited to extracting the morphological parameters of crops [92, 93], and parsing less phenotypic information, which is generally not used to extract physiological and biochemical parameters of crops. Zhang et al. applied on-board RGB images to effectively classify the crops and estimate the leaf area index in 6.5 km^2 crop planting area [94]. Hunt et al. obtained the images of corn, alfalfa and soybean in planting areas respectively through digital cameras, extracted DN values in green-light and red-light wavebands to construct normalized green minus red difference index (NGRDI), and then the crop biomass and nitrogen nutrient status, etc. were evaluated separately [95]. Zhou et al. used Canon EOS 5D

Mark III digital camera to acquire digital images of rice in critical growth periods, constructed the common color indexes and conducted the prediction of rice yield [96].

1.4.2 Multispectral Imaging Technique

In the research and practice of monitoring crop growth status with remote sensing technology, crop reflection spectrum is the most widely used spectral type, so the acquisition method and technology of crop reflection spectrum are the important basis for successfully monitoring crop growth monitoring [97, 98]. Multi-spectral remote sensing imaging refers to the technique of synchronous imaging for ground objects by using the sensors with two or more spectral channels, which can obtain crop radiation or reflection information in visible light bands(400–700 nm)-near infrared light bands (700–1200 nm). It is usually used to extract the plants biochemical parameters as pigment content (chlorophyll and carotene), leaf moisture content, and calculated vegetation index, and to estimate crop aboveground biomass, grain yield and the parameters related to photosynthesis [99, 100]. Li Yingxue et al. explored the quantitative relationship between nitrogen accumulation in wheat leaves and canopy reflectance spectral index [101]. Read et al. analyzed the relationship between nitrogen concentration and spectral ratio index of cotton (Gossypium hirsutum) leaves [102]. Xue Lihong et al. and Zhou Dongqin et al. used the spectral vegetation index combined with 810 and 1220 nm and visible light to monitor the nitrogen status of rice (Oryza sativa) plants [103, 104]. Petisco et al. applied the near-infrared spectroscopy to efficiently estimate nitrogen and cellulose content in leaves of different tree species [105]. Jiang and Ying applied the Fourier diffuse reflectance near-infrared spectroscopy to investigate the non-destructive monitoring method for chlorophyll content in Lapsicum annaum leaves, and the prediction correlation coefficient reached 0.975 [106]. Zhang Luda et al. established a near-infrared spectroscopy quantitative analysis model of different components in Nicotiana tabacum (total sugar and reducing sugar) and wheat (protein content) samples using principal component regression method and partial least square method (PLS), and the prediction performance of the model was good [107]. Cheng gan et al. found that there was a good correlation between the chlorophyll content in rice and the canopy spectral reflectivity at 640–690 nm, and the maximum correlation coefficient was around 680 nm [108].

Multi-spectral cameras are characterized by low price, flexibility, low weather impact, high level of operation and others. However, due to the low spectral resolution and less frequency bands of multispectral imaging, spectral reflection characteristics are susceptible to ambient light, and the disadvantages such as discontinuous spectrum [109], which severely limits its application in the study of crops phenotype.

1.4.3 Hyperspectral Imaging Technique

Hyperspectral imaging technology refers to the using the hyperspectral instruments to acquire a large number of very narrow and continuous spectral image data with high spectral resolution in a specific spectral domain, which can provide spectral information of tens to hundreds of narrow bands (usually the band width <10 nm) for each pixel, and produces a complete and continuous spectral curve. Hyperspectral images have the high spatial resolution and high spectral resolution, which can accurately reflect the spectral characteristics of crops in the field and the spectral differences between the crops [1–3, 110]. Using these hyperspectral images, crop canopy agronomic information, such as crop moisture content, chlorophyll content, LAI and other physical and chemical parameters, can be obtained more accurately, and then the material structure of target objects, chemical composition and complex biochemical processes of the leaf and canopy can be studied, thus further judge the attributes of the target object to facilitate prediction of crop growth and yield [111–113].

As a digital agricultural technology, hyperspectral imaging technology is widely used in the field of crop phenotype collection [114]. It can be used not only to detect a certain physiological indicator or a certain component content of crops, but also to observe certain organs of plants, and to identify some crop pests and diseases [115–118]. Zhou Dongqin et al. based on hyperspectral imaging technology to detect the carbon and nitrogen ratio of rice leaves, with the field experiments of different rice varieties and different nitrogen application levels, systematically analyzed the quantitative relation between carbon and nitrogen ratio and corresponding canopy hyperspectral reflection characteristics of rice leaves at different growth stages, and then determined the optimal monitoring model for carbon to nitrogen ratio in canopy leaves [4–6]. In the study on rice stem and leaf segmentation, Feng Hui et al. applied hyperspectral camera to collect rice spectral data, used principal component analysis to extract principal component images from the data, and then separated rice stems and leaves based on digital image processing technology to realize the non-destructive and accurate segmentation of rice stems and leaves [119]. In addition, hyperspectral imaging technology has also been successfully applied to the quality detection of food products, plant seed germination testing and other aspects [120, 121].

Hyperspectral imaging technology is one of the most promising imaging technologies in the field of crop phenotype research, but it is greatly affected by the field light environment and crop canopy structure, and its relatively miscellaneous and redundant information increases the speed and difficulty of hyperspectral image data analysis and processing [122]. Besides, the high price of hyperspectral imaging equipment also limits its promotion and use.

1.4.4 Infrared Thermal Imaging Technique

Infrared radiation is the electromagnetic wave radiation that is widely found in nature, all objects above absolute zero are constantly emitting infrared radiation [123, 124]. The sensitive bands of the thermal imager are 3–14 μm, the most commonly used bands are 3–5 and 7–14 μm, and in these two bands, the velocity of infrared radiation in air is close to its maximum value. Thermal imaging technology receives the infrared radiation emitted by the measured object through the infrared detector and the optical imaging objective lens, thereby generating an infrared thermal image to complete the visualization of the infrared radiation, which can visually present the surface temperature of the object and be used to study the characteristic information of the target object [125]. Meanwhile, the water-related indicators such as leaf water potential, stomatal conductance, transpiration efficiency and osmotic regulation have the significant correlation with canopy temperature, which can be used for monitoring crop growth under adverse conditions. In addition, under high temperature and drought conditions, lower canopy temperature has a significant positive correlation with yield [126, 127]. In studies of salt tolerance and heat tolerance, osmotic stress makes the stomatal closure, decreases the stomatal conductance, increases the leaf temperature, and reduces the photosynthesis, the changes in photosynthesis and stomatal conductance can be monitored through the spectrum reflectance from visible to near-infrared region [128]. Therefore, thermal imagers can also be used to detect the temperature rises caused by stomatal closure under conditions of high salt and water stress, in order to clarify the physiological responses (early thermal reactions) under plant stress conditions [129, 130], and to assess the extent of biotic or abiotic stress in crops [131]. Furthermore, with the rise of infrared non-destructive testing technology, thermal imaging technology has made a significant breakthrough in the early detection of crop diseases and pests [132]. Awad et al. [133] selected materials that were moderately sensitive to wheat powdery mildew, used infrared thermography to observe the leaf surface temperature of wheat, captured thermal infrared images and other relevant data, and compared a large number of thermal infrared images between the experimental group and the control group. It was found that the surface temperature of wheat leaves decreased slightly after 1 h infected with powdery mildew.

Although thermal imaging technology can quickly acquire crop transpiration status directly or directly, but its imaging system is susceptible to climate, especially in field's complex environment, environmental factors (such as background soil temperature, air temperature, wind and humidity, etc.) and the dynamic changes of crop canopy will affect the accurate measurement of the blade or canopy temperature [134, 135]. And this technique is using the difference of infrared radiation from different parts of the object surface for imaging, the partial temperature difference of some crops at room temperature is not obvious, so it is difficult to accurate imaging, and there are many different kinds of crops, a complete model of thermal radiation characteristic has not been constructed for thermal imaging image analysis at present [136]. Therefore, how to revise the thermal image is a key factor affecting its application in field phenotypic measurement and determination. In addition, the low resolution of the thermal imager will exacerbate the phenomenon of mixed pixels, thus increasing the deviation of the temperature measurement of the vegetation portion.

1.4.5 Fluorescence Imaging Technique

Fluorescence, a common luminescence phenomenon in nature, is produced by the interaction between photons and molecules, this mutual process can be described by the molecular level diagram of Jablonslc. When a substance with normal temperature is irradiated by incident light of a specific wavelength (usually ultraviolet or X-ray), it absorbs light energy and enters an excited state, and immediately retreats and emits the outgoing light with a longer wavelength than the incident light (usually in the visible light wavelength). Once the incident light stopped, the luminescence phenomenon disappears immediately. The outgoing light having this property is called fluorescence. Fluorescent sensors adopt a unique, active measurement method. Chlorophyll of crops is the part that produces fluorescence, the external light is absorbed by chlorophyll and used for photosynthesis, during which part of the absorbed light is converted into heat energy and the other part is returned in the form of fluorescence [137]. Therefore, some characteristics of crops, especially the parameters related to photosynthesis, can be obtained by detecting fluorescence. Chlorophyll fluorescence sensors can indirectly complete the detection of crops photosynthetic state, not only reflect the original reaction process of photosynthesis such as light energy absorption, excitation energy transfer and photochemical reaction, but also related with the establishment of the electron transfer and proton gradient, and with the process of ATP synthesis and CO2 fixation. It has wide applications in studying the influence of adversity (drought, high and low temperature, nutrient deficiency, pollution, disease, etc.) on crop photosynthesis [138].

In terms of crop phenotype collection, fluorescence imaging technology is mainly used for physiological responses to crop stress and detection of crop diseases and insect pests. Wang Chunping et al. studied the effects of low temperature stress on chlorophyll fluorescence characteristics of different leaf ages of rice seedlings, five different rice cultivars were selected for low temperature (4 °C) treatment, and the chlorophyll fluorescence parameters were measured by the modulation chlorophyll fluorescence imaging system [139]. Harbinson et al. applied this fluorescence imaging technique for high-throughput single-plant screening in crop breeding [140], Lee et al. used naphthalene fluorescent probes to detect the distribution of zinc in HeLa cells of arabidopsis thaliana [141], and Bauriegel et al. used fluorescent imaging technology for early disease detection of crops, et al. [142]. Although fluorescence imaging has certain advantages compared with other imaging technologies, the detection indicators are diverse, and it can detect the internal diseases or missing elements of crops, is not affected by environmental factors such as temperature and light when gathering information. However, fluorescence imaging technology can only detect chlorophyll or other physiological response parameters related to pigment content, mostly measured for single plant characteristics, and for large-area canopy information obtained by remote sensing platform, it is difficult to be used when analyzing phenotypic biological parameters, which limiting the scope of its application.

1.4.6 Radar Sensing Technique and Other Information Monitoring Technique

Accurate and rapid acquisition of spatial structure information of crop canopy plays an important role in detecting crop phenotypic biological information and in decision-making variable-rate fertilization. Along with the advancement of 3D information acquisition technology, 3D digitization technology can be used to accurately obtain spatial information of field crops in different scales [143, 144], such as Lidar sensing technology (LiADR), 3D reconstruction technology based on machine vision, ultrasonic positioning technology, time-of-flight techniques, often used to obtain the 3D spatial information of crop canopies [145–147]. Based on the monitored 3D information of crop canopy, the canopy structure parameters such as plant height, aboveground biomass, leaf orientation value, leaf area distribution, leaf height, leaf length and leaf width can be extracted [148–151] but when 3D imaging technology acquiring field crop phenotypic data, it is susceptible to crop canopy disturbances and lighting environment changes [152]. Of course, in addition to some common radar sensors for distance measurement, full-wave radar and hyperspectral radar are also applied in crop phenotypes: full-wave radar (FWF—Full-Wave-Form) not only records the peak intensity, but also preserves all the information returned by the laser pulse in a special waveform, making it easier to identify multiple objects. Hyperspectral radar adds crop response characteristics of multi-band spectral to point cloud data, which is conducive to diagnosing crop vigor [153, 154].

Radar sensing technology has been widely used in modern agricultural research since the 1980s. It is an active remote sensing technology that takes the laser as the emission light source and adopts photoelectric detection means, mostly based on Time-of-Flight (ToF), interferometry or triangulation principle, it can be divided into 2D radar and 3D radar according to the structure division. 2D radar detects the targets by sector scanning and generates planar point cloud data. 3D radar can acquire the high-precision, high-resolution point cloud data that characterizes the surface characteristics of the targets, which makes it easier to obtain the shape and structure data of the target. Advantages of high resolution, strong anti-interference ability and good low-altitude detection performance, make it able to quickly obtain the high-precision structure parameters of horizontal and vertical vegetation canopy. However, due to higher price and large amount of data processing, it is currently fresh and few applications in the study of crop phenotype, mostly applied to the analysis of tree biomass, plant height and other information [155].

Depth sensing technology is basically similar to the principle of the 3D laser sensor, except that 3D laser sensor is scanned point by point, while depth camera obtains the real-time depth information of the whole image at the same time, usually outputs depth, amplitude and intensity images [85]. The depth image contains Z-coordinate information, the amplitude image can evaluate the quality of the depth information, and the intensity image is a simple gray image. Current depth cameras can be divided into three types according to their working principle: structured light depth camera, TOF (Time of Flight) depth camera, RGB binocular depth camera. The main difference is that: the structured light camera has the advantages of mature technology, and the resolution of depth image can be done relatively high, but easily affected by illumination, basically can't be used outdoor environment. TOF camera has good anti-interference performance, is less affected by the environment, and has a wider view field, but the resolution of depth image is lower, which is not suitable for high-precision occasions. The cost of RGB binocular camera is the lowest compared with the previous two schemes, but the depth information is obtained relying on the pure software algorithm. The algorithm is of high complexity and difficulty, and processing chip requires high computational performance, so it is not applicable in the dim environment and the unobvious characteristics conditions. RGB-D camera is a newer, low-cost depth camera, equipped with RGB camera, depth sensor and infrared emitter, which can provide both color and depth information of the target object simultaneously, Kinect and Xtion are two commonly used depth cameras, which are applied in many fields such as crop measurement.

Microwave is a kind of electromagnetic wave, which refers to the electromagnetic wave with a frequency of 300 MHz to 300 GHz. It is the abbreviation of a limited frequency band in radio waves, has a higher frequency than ordinary radio waves, usually which is also called "UHF (ultra-high frequency) electromagnetic wave", with the characteristics of high frequency, large amount of information and good penetrability. Basic properties of microwaves usually appear as three characteristics of penetration, reflection and absorption. Microwave imaging mainly refers to an imaging method using microwave as an information carrier, whose essence belongs to the problem of electromagnetic inverse scattering. The imaging method is to irradiate microwaves into the measured target, make it pass the target and carry out the internal information, after receiving these information, a series of processing and operation in the computer are performed to form an image. Microwave technology is widely used in industry, aerospace, medical and other fields, in agriculture application, it is mainly used for three-dimensional imaging of individual plant, including 3D model study of plant type [156], root morphology detection [157, 158], and other aspects. Liu Rui et al. [159] obtained the point cloud data for each point of corn plants by performing the lidar scanning tests on corn, and then generated a polygon mesh or parametric curved surface slices through a series of algorithms, thus established the 3D model of maize plants. Microwave imaging technology is not affected by weather and distance, is also not limited by time and space, has strong penetrability to obstacles, and can be used for crop remote sensing recognition all day long [160]. However, its low imaging accuracy leads to the poor quality of acquired information, and the acquired cloud points are easy to be lost in the overlapping parts, which making the processing of imaging image difficult.

1.5 Real-Time Variable-Rate Fertilization System Based on Crop Phenotypic Biological Information

Based on the studies on high-throughput crop phenotypes conducted by many domestic and foreign scholars, along with the continuous development of computer science and remote sensing technology, and combined with the agronomic requirements of different types of crop cultivation, the detection platform of biological information parameters of field crop canopy has been more applied in real-time variable-rate fertilization. According to the canopy biological parameters obtained by the monitoring system in real time, the expert decision-making system of variable-rate fertilization set up previously is run to generate the target fertilizer requirement application online, guiding the core controller to adjust actuator and then to control the actual fertilizer application for the crop, which realize scientific and reasonable on-demand variable-rate input of fertilizer, so as to improve the fertilizer utilization efficiency and obtain higher crop yields, greater economic benefits, and better ecological environment. In accordance with the different application carriers, the monitoring platforms for crop phenotypic information are mainly divided into two types: aviation and ground (as shown in Fig. 1.2), among which the common aviation phenotypes monitoring technology is divided into satellite-borne and airborne ones, and the ground phenotypes monitoring technology can be divided into hand-held and vehicle-mounted ones [161].

The multi-source satellite images acquired by the satellite-borne platform can be used to extract crop phenotypic parameters, but due to the limited time and spatial resolution of data acquisition, it is difficult to be applied to small-area test regions and high-frequency dynamic monitoring [162]. The airborne phenotypic platform (helicopter or unmanned aerial vehicle (UAV)) has a larger load to meet multiple sensors at the same time, and can acquire multi-source image information of the entire test site in a short time [163–165], which is applied to measure the thousands of canopy temperature and lodging situation in field plots, as well as the change information on canopy structure at different test sites [166, 167]. Researchers at home and abroad have applied UAVs or drones to carry multi-spectral cameras (such as Mini-MCA6), hyperspectral cameras (such as Cubert UHD185) or active non-imaging sensors (such as RapidSCAN) to monitor the leaf area index, aboveground biomass and nitrogen nutrient status of crops such as wheat, rice, cotton and corn etc., [168–170] and all of which achieved good results. For example, Japan used TM remote sensing data to monitor nitrogen, amylose, amylopectin and other quality indicators of rice, and guided the rational application of nitrogen fertilizer, so that the nitrogen content of rice grain in the larger experimental area was reduced from 7.7% to 7.3%, which improved the rice quality level and brought significant economic benefits [171]. However, the airborne platform costs more, requires more

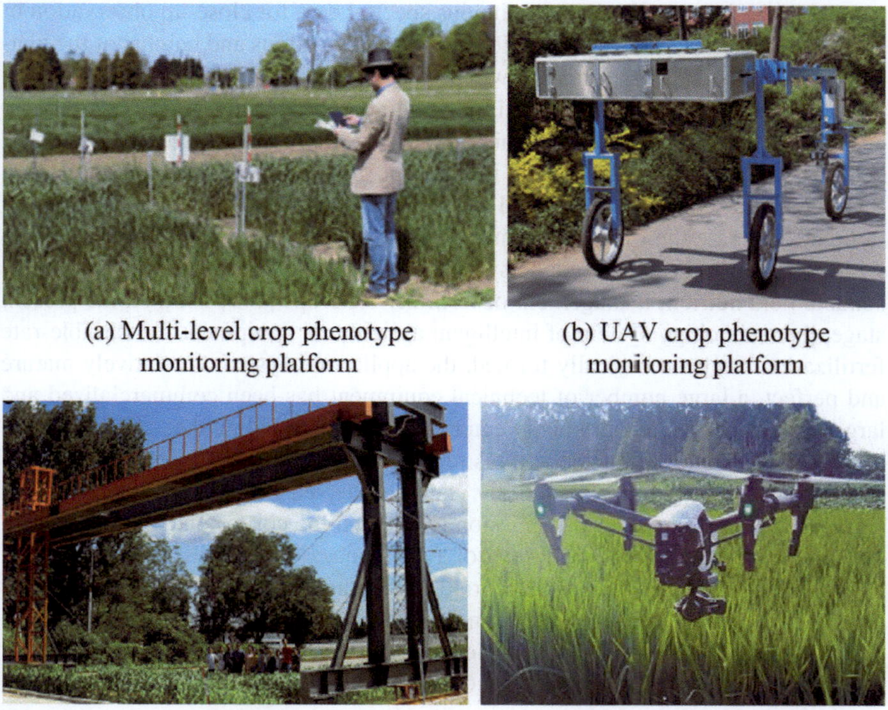

(a) Multi-level crop phenotype monitoring platform

(b) UAV crop phenotype monitoring platform

(c) IoT distributed phenotype monitoring platform

(d) Mobile vehicle-mounted phenotype monitoring platform

Fig. 1.2 Phenotyping monitoring platform for field crop. (**a**) Multi-level crop phenotype monitoring platform. (**b**) UAV crop phenotype monitoring platform. (**c**) IoT distributed phenotype monitoring platform. (**d**) Mobile vehicle-mounted phenotype monitoring platform

personnel maintenance, and the remote sensing information obtained by each sensor is limited, so multi-sensor coordination and data fusion are required. Moreover, the canopy disturbance is relatively large when acquiring high-resolution images at low altitude [172–174]. The ground phenotypic platform is usually using the common agricultural equipment equipped with a series of phenotype sensors to approach the crop canopy to obtain phenotype information, and GPS is installed for positioning and navigation [175, 176]. The ground phenotypic platform can simultaneously acquire multi-source information (such as multi-spectral, hyperspectral and thermal infrared images), since different sensor groups differ greatly under different test points and environmental conditions, so how to calibrate the sensor groups to ensure the reliability and comparability of multi-region data is the main technical problem of the vehicle phenotype platform [177].

At present, the phenotypic biological information data acquisition technology of field crop, which is widely applied in the real-time variable-rate fertilization field of precision agriculture, is still mostly concentrated on the ground vehicle-mounted operation. The vehicle-mounted phenotype omics platform is simple to operate and can obtain high-resolution and continuous spectral data for close-up observation of crop phenotypes. Sensors are mounted at different locations and heights of the platform depending on the type and growth period of the crop [178], related accessories include power system, data acquisition terminal, GPS receiver, encoder, etc., to ensure the running work of the vehicle system. A high-performance terminal that can receive various sensor data is needed and important, equipped with Beidou Satellite, radar, GPS, gyroscope and other sensors, it can realize automatic navigation and generate the distribution map of crop phenotype feature, and coordinate with the optimized expert decision system of variable-rate fertilizer to conduct reasonable field nutrient management of fertilizer for crops [179, 180]. At the present stage, the technology system of intelligent agricultural equipment for variable-rate fertilization has been basically formed, the application system is relatively mature and perfect, a large number of technical equipment has been commercialized and large-scaled applied, and gradually entered the international market.

Professor Qamar U. Zaman and his research group in Canada based on the biological data information of blueberry canopy detected by color camera, through wireless Bluetooth module transmission to respond the corresponding control procedures, adjust the variable-rate input of a single nozzle for fertilizer consumption, to achieve efficient, reliable and accurate fixed-point variable-rate fertilization for wild blueberries, improving farm economic benefits [181] and ecological benefits [7] (as shown in Fig. 1.3). Karl-Heinz Dammer et al., Leibniz Agricultural Engineering Association, Potsdam, Germany, developed a fungicide variable

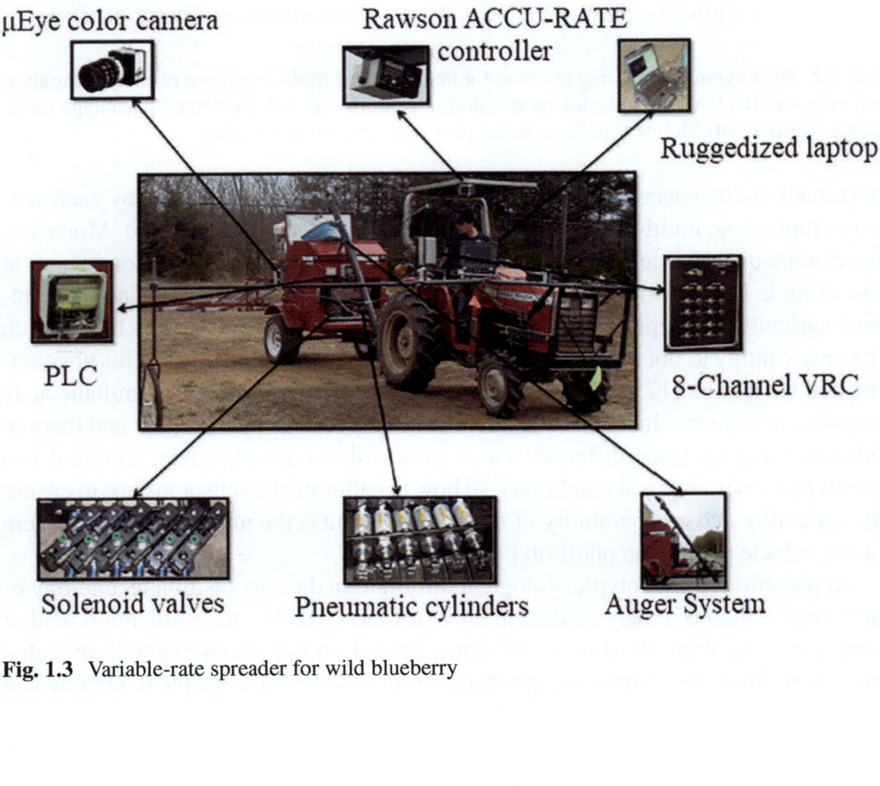

Fig. 1.3 Variable-rate spreader for wild blueberry

sprayer based on the leaf area index (LAI) of cereal crops, as shown in Fig. 1.4. The crop biomass density is measured in real time by a portable spectral sensor (CROP-Meter) fixed at the front of the tractor, matching the numerical model of leaf area index value of crop canopy, and the GPS positioning information was combined to output the control signal of sterilization dose, so as to drive the hydraulic valve port to adjust the fungicide flow rate. Thereby to control variable-rate spraying [182].

Professor D. EHLERT's team developed a variable-rate spreader of nitrogen fertilizer for winter wheat based on the crop density detector with mechanical pendulum (as shown in Fig. 1.5), the onboard computer continuously adjusts the fertilizer application of the spreader based on the crop density parameters measured at the

Fig. 1.4 Variable-rate fungicide sprayer

Fig. 1.5 Variable-rate nitrogen fertilizer of winter wheat

fixed point. Multiple field trials on calcium ammonium nitrate fertilizer (nitrogen fertilizer) in different operating seasons showed that the fixed-point variable-rate fertilization by this spreader could save 10–12% of fertilizer application and improve the crop yield and quality [183]. At the University of Tennessee, in order to improve the utilization rate of nitrogen fertilizer during the cotton production process, and to optimize the impact of soil reflection during the GreenSeeker sensor measurement process, Mariso [184] designed the related optimization algorithm to achieve the precise application of N-fertilizer, and the ultrasonic device was developed to ensure that the GreenSeeker sensor is always 36 inches from the cotton canopy, as shown in Fig. 1.6.

A maize variable fertilization machine based on spectral detection technology introduced by Professor Stephen from Texas A&M University, as shown in Fig. 1.7. Spectrum monitoring mechanism, including a halogen lamp and an N sensor, was installed behind the tractor to monitor the nitrogen nutrition status of the corn in real time through the N sensor, and using it to guide the field fertilization operation [185]. Researchers at the University of Paris XI used GreenSeeker sensors and Multiplex sensors to monitor the nutritional status of the grapes in real time and to guide field fertilizer management, as shown in Fig. 1.8. This machine can obtain various field indices such as the Chlorophyll Fluorescence Index (SFR), Nitrogen Balance Index (NBI) and Flavonoid Index (FLAV) through the Multiplex sensors [186].

In addition, in order to study and compare the performance of different spectral sensors, Kenneth from the University of Missouri installed three kinds of sensors of GreenSeeker, Crop Circle and Crop Spec on the same fertilizer machine, as shown in Fig. 1.9. Their test results showed that the performance of the Crop Spec is

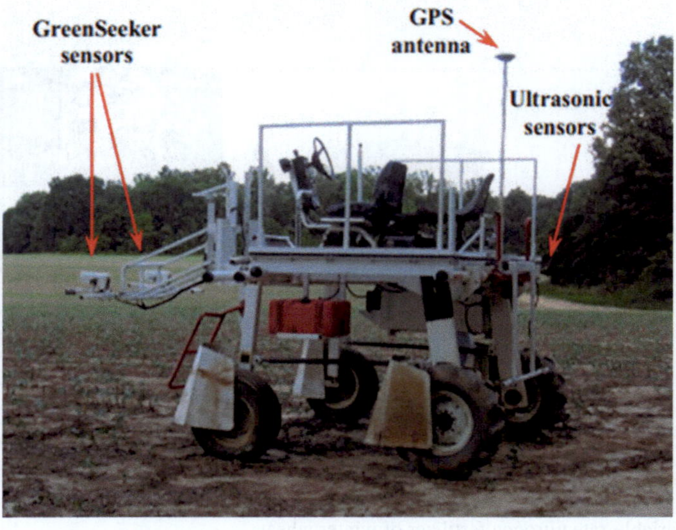

Fig. 1.6 Cotton variable-rate fertilizer applicator

(a) Back view of the sprayer. (b) Closer view of the lighting chamber.

(c) Inside of the lighting chamber: front view. (d) Inside of the lighting chamber: side view.

Fig. 1.7 Corn variable-rate fertilizer applicator. (**a**) Back view of the sprayer. (**b**) Closer view of the lighting chamber. (**c**) Inside of the lighting chamber: front view. (**d**) Inside of the lighting chamber: side view

superior to the other two sensors, and the other two sensors are greatly affected by the position. Due to the uneven terrain in the field, the measurement results fluctuated greatly with sensors other than Crop Spec [187].

China's research on precision agricultural production and variable-rate fertilization technology started relatively late, the real-time variable-rate fertilization research based on crop phenotypic biological information is still in its infancy. It is mainly through the introduction of foreign advanced agricultural production machinery and the supporting control systems, the digestion and absorption of foreign technology is the mainstay, and there is a lack of complete systematic theoretical research and related experiments [188]. In recent years, with the continuous development and improvement of the relevant national policies 'Application Reduction and Efficiency Increase' and variable-rate technical systems, the experts and scholars from Chinese universities and research institutes in the field of agricultural machinery intelligent control have independently innovated on the basis of

Fig. 1.8 Grape variable-rate fertilizer applicator

Fig. 1.9 Variable-rate fertilizer applicator with multiple sensor

absorbing experience of foreign variable-rate technology, groping and exploring in the variable-rate fertilization of precision agricultural in succession, and developed a variety of variable-rate fertilization equipment adapted to the national conditions of China characteristics.

Professor Ma Xu's research team from the College of Biological and Agricultural Engineering of Jilin University used the MSR16R multispectral instrument to detect canopy spectral reflectance information of soybean crops, established a nutrition diagnosis model in soybean mature stage to estimate the nitrogen content required for growth, and the expert decision-making system for variable-rate fertilization was determined using fuzzy control theory, then developed a precision variable-rate liquid fertilizer applicator for soybean, as shown in Fig. 1.10, and the field variable-rate spray for liquid fertilizer has certain stability [189]. Meng Zhijun and Zhao Chunjiang research team, from National Engineering Research Center for Information Technology in Agriculture, developed an online measurement variable-rate fertilizer applicator based on crop growth (as shown in Fig. 1.11) [190], using non-destructive photoelectric inspection technology, through normalized difference vegetation index (NDVI) measuring instrument to obtains the NDVI in real time, and the core processor controlled the different fertilizer application according to the fuzzy control algorithm. It can realize according to the real-time changes of crop growth information to conduct online living diagnosis, through the correlation between crop canopy spectral reflectance and growth, in the form of active light source for sensor monitoring, access to obtain the growth information of community growth trend and leaf age index during the whole growth period of crops. Interest, less fertilization in areas where crop growth is better, and appropriate more fertilization in areas with poor growth, so as to realize precision variable-rate fertilization according to crop growth.

Wang Xiaochang and his research team from Nanjing Agricultural University determined the demand situation of chemical fertilizer (nitrogen fertilizer) based on

Fig. 1.10 Variable-rate fertilizer for soybean

Fig. 1.11 Variable fertilizer machine based on crop growth

Fig. 1.12 Variable fertilizer based on canopy spectral reflectance

the NDVI value (normalized vegetation index) of winter wheat canopy monitored in real time, and implemented the intelligent control program of variable-rate fertilization to complete the collection of relevant information parameters for variable-rate fertilization operations (as shown Fig. 1.12), at the same time, decision-making generated the real-time target topdressing application, realized the online adjustment of the fertilization amount, and achieved the goal of precise variable-rate fertilization. Furthermore, Shi Yinyan et al. in this team developed a centrifugal and high-efficiency variable-rate fertilizer spreader based on real-time growth information of rice (as shown in Fig. 1.13). Greenseeker spectrum detection system obtained the DNVI values of the rice canopy in real time, run the expert decision-making system of variable-rate fertilization previously set up in operation, based on the optimized Raun model to generate the target fertilizer requirement application in real time, and the real-time spreading application was adjusted online to realize the real-time accurate variable-rate fertilizer spreading for rice in relative sense.

It can be seen that the real-time variable-rate fertilization technology based on the vehicle-mounted phenotype omics platform has great application potential in the research of crop phenotype bioinformation monitoring and variable-rate fertilization technology research of precision agricultural.

Fig. 1.13 Variable spreader in spectral reflectance for rice

Moreover, of course, other vehicles that can be used to carry sensors and high-throughput phenotypic information for field crops include airships, parachutes, hydrogen balloons, and ground observation towers, etc., but are limited by their flexibility and cost, such as line spacing, plant spacing and crop height limitations, make it difficult to measure different crops or different growth periods, especially at maturity, and their image resolution, safety performance and load capacity all need to be improved, so these platforms are usually less applied in the field of variable-rate fertilization.

1.6 Existing Problems and Suggestions for Real-Time Variable-Rate Fertilization Technology Based on Crop Biological Information

In recent years, through the continuous efforts of scientific research personnel from agricultural universities and research institutes in China, studies on crop phenotype information monitoring and PA variable-rate fertilization technology have made some progress, but in the long run, the real-time variable-rate fertilizer technology and its application based on crop phenotype bioinformation are still in the experimental research stage, and still exist certain problems, mainly manifested in the following aspects:

1. Study object, monitoring area and scope are relatively limited. At present, the research on variable-rate fertilization technology based on crop phenotypic biological information mainly focuses on the field conditions regions of a few specific varieties of crops where with relatively developed economy, larger planting area and relatively strong research ability. It has obvious regional features and uneven distribution, lacks in-depth research on phenotypic information acquisition and target recommended fertilization for multi-species crops in agricultural

underdeveloped areas, and mainly targets wheat, corn, rice and barley and other cereal crops, but for beans, potatoes and other characteristic commercial crop, there have been few research and rarely reported.

2. Obtained biological parameters of crop phenotypes information are single. Various imaging techniques have their own restriction conditions, affected by the external temperature, illumination and other complex field environment factors. Some remote sensing imaging techniques are limited by their measurement accuracy, the spatial resolution and spectral resolution is low and the measurement error is large, which cannot be widely applied. It is necessary to explore the fusion of different phenotypic data sources at different stages of data processing, overcome or reduce the error of detection results brought by a single data source, improve the accuracy of crop phenotypic parameters extraction and the stability of variable-rate fertilization equipment application.
3. Rapid processing of the monitored enormous remote sensing data faces huge challenges. Imaging hyperspectral instrument and lidar sensors quickly acquire a large number of high-temporal resolution remote sensing information, which is limited by computer hardware performance and data processing methods. It is still difficult to realize big data online rapid processing in real time [11], and the mixed spectrum of crops in different growth stages with the composition soils has a complex mechanism, the mixed spectrum of the field components is difficult to decompose, resulting in a reduction in the accuracy of target fertilizer requirement with inversion calculation.
4. Constructing inversion model of fertilizer requirement application based on remote sensing information is lack of common methods. In practical application, when parsing the phenotypic information for different crops and different growth stages needs to be modeled separately, ignoring the mutual influence of other physiological parameters interplay and comprehensive effects, and the application of the model have environmental conditions limitations, variability and uncertainty, including the uncertainty from the driving variables or parameters of the remote sensing inversion model and the uncertainty caused by the error propagation of the model itself, leading to crop growth model and nutrient recommendation system cannot be closely matched with the precision variable-rate fertilization machine, and the update cycle of fertilization expert decision analysis system become slow down.
5. Intelligent control system has quite weak independent research and development capabilities. Various types of sensors have high cost, the vehicle-mounted hyperspectral, laser radar and chlorophyll fluorescence sensors are basically dependent on imported products, the price is so high. The core technology of them cannot be targeted optimized and upgraded, resulting in the system ability to accurately acquire and quickly respond to information cannot meet the practical operational demands, and lack the detection sensors of crop phenotypic biological parameter with independent intellectual property rights and can meet the needs of crop cultivation in China, to improve the response speed of real-time variable-rate fertilization and fertilizer precision, which restricting the extensive

promotion and application of the variable-rate fertilization equipment based on crop remote sensing phenotypic information in modern planting areas in China.

1.7 Development Trend and Prospect for Real-Time Variable-Rate Fertilization Technology Based on Crop Biological Information

Real-time variable-rate fertilization technology based on crop phenotypic biological information, as an emerging implementation method of precision agriculture, is an inevitable trend of intelligent precise agricultural development in the future. It has begun to take shape in foreign developed countries, but is still in the trial exploration stage in contemporary China, it should be further researched and in-depth promoted, especially for high-throughput phenotypic information monitoring of crops and the integration and application of real-time variable system, and more technical research breakthrough and collaboration are needed. On the basis of learning from the advanced experience of foreign countries, combined with the agricultural machinery and agronomic requirements of actual production of crop planting in China, the matching variable-rate fertilizer equipment with independent innovation should be studied and developed, reducing the production costs, can be widely demonstrated and promoted in modern agricultural planting areas, thereby realizing preferably effective and efficient utilization of fertilizer resources to achieve healthy and sustainable development of green ecological agriculture. Therefore, in the future, we believe that real-time variable-rate fertilization techniques concerning crop phenotypic biological information will continue to receive attention, and its main development or conquering direction includes:

1.7.1 Continuously Expand the Breadth and Depth of Crop Phenotypes, and Strive for the Extraction of High-Latitude Parameters and Multi-scale Analysis

Field crop phenotype biological information acquisition based on remote sensing technology has significant advantages and broad prospects in the application of real-time variable-rate fertilization of precision agriculture, its flexibility, timeliness and high efficiency are the important characteristics necessary for crop phenotype information extraction and real-time variable-rate fertilization. However, the current research object on the crop phenotypes biological information with field vehicle-mounted crop phenotypes mainly focuses on the specific varieties of common grain crops such as rice, wheat, corn and barley, and there are few studies on the characteristic commercial crops like beans and potatoes, and most studies are only devoted to extracting or parsing single-scale phenotypic biological parameters, lacking crop phenotypic information identification and fertilizer requirements analysis of target crop based on remote sensing technology in complex farmland environment. And the accurate description of the intrinsic nutrient status during crop growth depends on the high-dimensional characteristic data obtained by the detection sensor and the data analysis of different scales, as the extraction of individual plant information from the population and specific traits of organs (such as root system and flower spike) from individual plant. Therefore, it is necessary to continuously expand the breadth and depth of studies on parsing crop phenotypes by remote sensing and devote to the extraction and multi-scale analysis of high-latitude parameters, which will lay a theoretical foundation and technical reference for promoting the application of crop phenotype biological information extraction using near-earth remote sensing technology applied in precision agriculture real-time variable-rate fertilization technology.

1.7.2 Deeply Mine and Integrate Multi-source Remote Sensing Data to Construct Analytic Model of High-Throughput Phenotypic Information and Recommendation Model of Target Fertilizer Demand Application

A large number of narrow-band and continuous crop hyperspectral images obtained by remote sensing technology can more comprehensively present the unique spectral characteristics of crops and accurately obtain the biophysical and biochemical status of crops. And the hyperspectral data contains abundant information and large amount of data processing, at present, it still does not fully play an important role in crop precision recognition and growth state monitoring. It is necessary to further strengthen the mechanism research of using hyperspectral remote sensing to analyze crop phenotypic information and deeply dig into effective information, achieving the high-precision analysis of crop three-dimensional size, biomass, vegetation index and other indicators. Moreover, most of the analysis models for field crop phenotypic information and recommendation models for target fertilizer demand application constructed based on a single method have the disadvantages of poor versatility and poor interannual prediction stability, and are susceptible to crop planting areas, crop variety types, and crop growth periods. So in order to increase the prediction accuracy of crop phenotypic information and the inversion model of intrinsic nutrient, it is necessary to combine multi-source remote sensing information (such as color data, depth data, spectral data, etc.), environment and crop physiology knowledge, and apply machine learning or deep learning algorithms in the pre- and post-processing stages of high-throughput phenotypic data, using a variety of methods to construct strong versatility, great stability and high precision analysis models for high-throughput crop phenotypic information and recommendation models for target fertilizer requirements application, and the systematic test and verification are then conducted.

1.7.3 Persistently Accelerate the Development of High-Resolution, Low-Cost Sensors with Independent Intellectual Property Rights Adapted to China's National Conditions

High cost and low resolution (spatial resolution, spectral resolution) are the bottleneck factors that restrict the promotion and application of remote sensing technology in crop phenotypic information detection and real-time variable-rate fertilization operations. At present, most of the sensors, hyperspectral instruments, thermal imaging cameras and other equipments carried on vehicle-mounted remote sensing platform in China are mostly imported products from Europe or the United States, the price is higher, and the parsing of crop growth status has great errors with the growth of different crop varieties in various regions of China. And the key core technologies cannot be optimized and upgraded according to the actual growth process of crops in China, which seriously restricts its application in the study of crop phenotypic information parsing and real-time variable-rate fertilization technology. Therefore, in the course of future development, it should implement comprehensive strategy based on the actual planting characteristics, growth characteristics and management characteristics of different regions, different crops and different varieties in China, combined with scientific and rational fertilizer application methods and proportions of various regions and crops, so as to research and development the detection sensors with independent innovation ability for crop phenotypic biological information in line with the usage characteristics of Chinese fertilizers, which is applied to improve the response speed and discharge accuracy of control system for variable-rate fertilization, and effectively promote the application of remote sensing technology in phenotypic biological information monitoring process for crops, which provides infrastructure for real-time variable-rate fertilization technology of precision agricultural and promote green, healthy, sustainable development of intelligent precision agriculture.

1.7.4 Effectively Promote Multi-disciplinary, Multi-field Intersection Data Sharing Platform for Crop Phenotype and Fertilizing Decision

Research of crop phenotypes and precise variable-rate fertilization concerned to the fields of plant science, information technology, biotechnology, agricultural technology and engineering technology, is a composite science and technology with multi-disciplinary cross and multi-field integration, requiring deep joint multi-system, interdisciplinary, authoritative expert knowledge and auxiliary decision-making knowledge, according to the nutrient demand mechanism and balanced fertilization principle of crop different growth period, combined with the advanced technologies such as network transmission, information perception and data processing, to carry out the study on high-precision and comprehensive coverage efficient detection and collection technology of soil and crop nutrient information, as well as on digital nutrition decision management system, truly integrate the agricultural engineering discipline with the basic theory discipline. And based on the effective nutrient requirements of crops, plant growth, water conditions, visible symptom diagnosis, target yield, fertilizer methods and other indicators, such as to establish a set of database sharing platform of crop phenotype and fertilization decision with multi-disciplinary cross and multi-field joint, to effectively promote the deep cooperation between interdisciplinary subjects, feasibly transformed technology into productivity, and jointly promoted the development of precision agriculture in China.

1.7.5 Profoundly Intensify Efforts to Cultivate Technical and Practical Talents for Phenotypes Detection and Precision Agriculture, and Popularize Variable-Rate Fertilization Technology with Phenotypes Information

The massive data of digital image, spectral imaging and thermal imaging that can be obtained by the remote sensing platform need to undergo a series of professional processing processes such as geometric calibration, radiation correction and data modeling, and finally realize the remote sensing means to analyze crop phenotype information, which requires practitioners has high scientific and cultural quality and the ability to use scientific computers and intelligent mechanical equipment. However, at present, end users lack the processing skills for the above data, and it is difficult to adapt to their active and effective development. Therefore, it is necessary to strengthen the cultivation of agricultural scientific and technological talents and the financial support for agricultural science researchers, and intensify efforts to cultivate high-quality precision agricultural applied talents. The training of agricultural science and technology personnel, especially young and middle-aged farmers, is the important measure for actively promoting crop phenotypes research, healthy development of precision agriculture, and meeting the new challenges of the century, is also the important issues that whether precision agriculture can be accepted and mastered by the majority of farmers. At the same time, it is necessary to speed up the development and demonstration application of the full-process technical solutions that are easy to operate, and integrate different functional software, so that enable users can independently realize crop growth monitoring, yield prediction and fertilization recommendation, etc., providing theoretical support for the promotion of real-time variable-rate fertilization technology based on crop phenotypic biological information nationwide in China.

1.8 Conclusion

Since the concept of precision agriculture was advocated and put into practice by American agricultural producers in the 1980s, variable-rate fertilization technology has developed rapidly in the field of intelligent agricultural equipment innovation in foreign developed countries. In recent years, crop phenotypic biological information based on agricultural remote sensing has made significant progress in the fields of crop area, growth, disaster, yield, quality and other monitoring technologies, the technical superiority in supporting the management of agricultural fertilizer, water, medicine, and damage have gradually been obtained industry recognition, and accelerates the rapid and sustainable development of precision agriculture real-time variable-rate fertilization technology. China's crop phenotypic biological information monitoring technology and real-time variable-rate fertilization technology based on agricultural remote sensing have changed from the phase of studying abroad valuable experience to a new stage of developing specific precision agricultural variable-rate technology with Chinese characteristics and in combination with China's national conditions, and made a series of creative achievements of research and application demonstration, which has laid the development advantage of China as a world superpower agricultural country studying in the field of crop phenotype monitoring and precision variable-rate fertilization technology. However, compared with other developed countries, there are still certain problems and gaps in some important key core technologies.

Crop models based on remote sensing technology are recognized as important new methods in modern agricultural research, which can digitize agricultural processes and raise the agricultural science from experience level to theoretical level. In the research process of high-throughput crop phenotypic biological information, it is urgent need to convert massive observation data into effective information for agricultural monitoring, and to construct the analytical methods and decision-making models of information acquisition, correction and fusion throughout the whole growth periods of crop, focusing on the study of 'satellite-vehicle-ground' multi-platforms collaborative synergy for farmland parameters under the large-scale planting conditions, and cooperating with precision acquisition technology, precision decision-making technology, and precision fertilization technology (variable-rate fertilization technology), to improve the independent innovation ability and technical reserve level, promote the application of theoretical research achievement transformed into practice, and enhance the leading ability of agricultural information technology to agricultural industry development. And the real-time variable-rate fertilization technology based on crop phenotypic biological information is absolutely an important development direction for realizing intelligent precision agriculture in the future, which will surely have a huge boost impact on promoting the sustainable development of China's agricultural modernization.

Acknowledgments The authors acknowledge the financial support provided by National Natural Science Foundation of China (Grant No. 51905281), and National Key Project of Research and Development Program (Grant No. 2016YFD0200600-2016YFD0200602-4). The authors would like to thank the teacher's and supervisor's technical support. We also appreciate the assistance provided by brothers and sisters during the tests, and we would like to thank Editage for providing English language editing. We are also grateful to the editor and anonymous reviewers for providing helpful suggestions to improve the quality of the present paper.

References

1. Pengcheng H (2018) High-throughput field morphological phenotyping using UAV-based proximal sensing. China Agricultural University, Beijing
2. Rui P, Qinxue X, Wenying Z (2016) Research advances of digital image technology and its application in crop phenotype. J Yangtze Univ (Nat Sci Ed) 13(21):38–41, 46
3. Fiorani F, Schurr U (2013) Future scenarios for plant phenotyping. Annu Rev Plant Biol 64(4):267–291
4. Dongqin Z, Yan Z, Jie Y, Yongchao T, Xia Y, Weixing C (2009) C/N content ratio of rice leaf monitoring based on canopy hyperspectral parameters. Trans Chin Soc Agric Eng 25(3):135–141
5. Dongqin Z, Yongchao T, Xia Y, Yan Z, Weixing C (2008) Quantitative relationships between leaf total nitrogen concentration and canopy reflectance spectra of rice. Chin J Appl Ecol 19(2):337–344
6. Jie Y, Yongchao T, Xia Y, Weixing C, Yusen Z, Yan Z (2009) Hyperspectral estimation model for chlorophyll concentrations in top leaves of rice. Acta Ecol Sin 29(12):6561–6571
7. Chattha HS, Zaman QU, Chang YK, Read S, Schumann AW, Brewster GR, Farooque AA (2014) Variable rate spreader for real-time spot-application of granular fertilizer in wild blueberry. Comput Electron Agric 100(1):70–78
8. Baohua L, Xinping C, Zhenling C, Qingfeng M, Ming Z (2015) Research advance in yield potential and yield gap of three major cereal crops. Chin J Eco-Agric 23(5):525–534
9. Zhenhua X, Caijuan G, Wenqi M, Jing W (2011) Study on the relationship among yield, nutrient efficiency and benefits of grain crop in typical regional areas of China. Chin Agric Sci Bull 27(11):116–122
10. Yuan J, Liu CL, Li YM, Zeng Q, Zha XF (2010) Gaussian processes based bivariate control parameters optimization of variable-rate granular fertilizer applicator. Comput Electron Agric 70(1):33–41
11. Jiuxin G, Yali K, Kailiu X, Donghai L, Xumeng F, Ning L, Min W, Shiwei G (2016) Effects of nutrient management on yield and nitrogen use efficiency of direct seeding rice. Acta Agron Sin 42(7):1016–1025
12. Bierman PM, Rosen CJ, Venterea RT, Lamb JA (2012) Survey of nitrogen fertilizer use on corn in Minnesota. Agric Syst 109(6):43–52
13. Yu H, Liu D, Chen G, Wan B, Wang S, Yang B (2010) A neural network ensemble method for precision fertilization modeling. Math Comput Model 51(11–12):1375–1382
14. Chunjiang Z (2009) Research and practice of precision agriculture. Science Press, Beijing
15. Jin Y, Qinhua L, Xuemei L, Teng Z, Xiaohui Z (2014) Granular multi-flows fertilization process simulation and tube structure optimization in nutrient proportion of variable rate fertilization. Trans Chin Soc Agric Mach 45(11):81–87
16. Yuxue G, Jin Y, Chengliang L (2011) FIS-based method to generate bivariate control parameters regulation sequence for fertilization. Trans Chin Soc Agric Eng 27(11):134–139

17. Rui L, Yongqian D, Hongfeng Y, Yingying D, Jing Y, Hongxu S, Weimin D (2017) Construction and application of the electronic prescription map system of variable rate fertilization in Southern Jiangsu. J China Agric Univ 22(8):124–132
18. Zaman QU, Esau TJ, Schumann AW, Percival DC, Chang YK, Read SM, Farooque AA (2011) Development of prototype automated variable rate sprayer for real-time spot-application of agrochemicals in wild blueberry fields. Comput Electron Agric 76(2):175–182
19. Han T, Jinwu W, Changsu X, Wenqi Z, Jinfeng W, Xiu W (2019) Research progress analysis on key technology of chemical fertilizer reduction and efficiency increase. Trans Chin Soc Agric Mach 50(4):1–19
20. Raun WR, Solie JB, Martin KL, Freeman KW, Stone ML, Johnson GV, Mullen RW (2005) Growth stage, development, and spatial variability in corn evaluated using optical sensor readings. J Plant Nutr 28(1):173–182
21. Jie C, Liang T, Xiaojun L, Weixing C, Yan Z (2011) Modeling plant nitrogen uptake and grain protein accumulation in rice. Sci Agric Sin 44(10):1997–2004
22. Zhongcai X, Huiyuan G, Jie L, Tangyuan N, Shenzhong T, Yu W (2011) Monitoring growth and grain yield of wheat in fields with different soil fertility levels and different fertilizer application using spectral reflectance technique. J Triticeae Crops 31(2):324–330
23. Maleki MR, Mouazen AM, Ramon H, De Baerdemaeker J (2007) Multiplicative scatter correction during on-line measurement with near infrared spectroscopy. Biosyst Eng 96(3):427–433
24. Zhao Y, Zhang Y, Han Y, Yu X, Li L, Chen W, Tian Y (2010). Assimilation technique of remote sensing information and rice growth model based on particle swarm optimization. Journal of Remote Sensing 14(6):1233–1249
25. Yabin M, Xiaoyu S, Bin Q, Guijun Y, Xia J, Xiao L (2015) Temporal and spatial variation of plant nitrogen and biomass of winter wheat under variable rate fertilization conditions. J Triticeae Crops 35(11):1576–1587
26. Xue L, Yang L (2008) Recommendations for nitrogen fertiliser topdressing rates in rice using canopy reflectance spectra. Biosyst Eng 100(4):524–534
27. Huajun T (2018) Progress and prospect of agricultural remote sensing research. J Agric 8(1):167–171
28. Zhongxin C, Jianqiang R, Huajun T, Yun S, Pei L, Jia L, Liming W, Wenbin W, Yanmin Y, Hasiyuya (2016) Progress and perspectives on agricultural remote sensing research and applications in China. J Remote Sens 20(5):748–767
29. Haboudane D, Miller JR, Tremblay N, Zarco-Tejada PJ, Dextraze L (2002) Integrated narrow-band vegetation indices for prediction of crop chlorophyll content for application to precision agriculture. Remote Sens Environ 81(2–3):416–426
30. Yubin L, Xiaoling D, Guoliang Z (2019) Advances in diagnosis of crop diseases, pests and weeds by UAV remote sensing. Smart Agric 1(2):1–19
31. Basso B, Ritchie JT, Cammarano D, Sartori L (2011) A strategic and tactical management approach to select optimal N fertilizer rates for wheat in a spatially variable field. Eur J Agron 35(4):215–222
32. Xiaoyu S, Jihua W, Wenjiang H, Guangjian Y, Hong C (2009) Monitoring spatial variance of winter wheat growth and grain quality under variable-rate fertilization conditions by remote sensing data. Trans Chin Soc Agric Eng 25(9):155–162
33. Yinyan S, Man C, Xiaochan W, Odhiambo MO, Weimin D (2018) Numerical simulation of spreading performance and distribution pattern of centrifugal variable-rate fertilizer applicator based on DEM software. Comput Electron Agric 144(1):249–259
34. Jianning Y, Jingchao Z, Xia Y, Yan Z, Weixing C, Jun N (2013) Experiments on performance of portable plant growth monitoring diagnostic instrument. Trans Chin Soc Agric Mach 44(4):208–212, 201
35. Xiaochan W, Man C, Guoxiang S, Yu Z, Yongnian Z (2015) Design and test of control system on variable fertilizer applicator for winter wheat. Trans Chin Soc Agric Eng 31(Suppl 2):88–92

36. Yinyan S, Zhichao H, Xiaochan W, Odhiambo MO, Weimin D (2018) Motion analysis and system response of fertilizer feed apparatus for paddy variable-rate fertilizer spreader. Comput Electron Agric 153(10):239–247
37. Yinyan S, Man C, Xiaochan W, Morice OO, Chengguang L, Weimin D (2018) Design and experiment of variable-rate fertilizer spreader with centrifugal distribution cover for rice paddy surface fertilization. Trans Chin Soc Agric Mach 49(3):86–93+113
38. Shi Y, Chen M, Wang X, Li C, Odhiambo MO, Ding W (2018) Numerical simulation and optimization of scattering performance of a conical centrifugal variable rate fertilizer spreader. Int Agric Eng J 27(1):34–43
39. Shi Y, Zhichao H, Wang X, Odhiambo MO, Sun G (2018) Fertilization strategy and application model using a centrifugal variable-rate fertilizer spreader. Int J Agric Biol Eng 11(6):41–48
40. Zhou S, Zongzheng L, Yuanyuan Y, Yan G (2015) Status and prospect of agricultural remote sensing. Trans Chin Soc Agric Mach 46(2):247–260
41. Chunjiang Z (2014) Advances of research and application in remote sensing for agriculture. Trans Chin Soc Agric Mach 45(12):277–293
42. Qiong W, Bo Q, Tuanjie Z, Xinfeng Y, Yan Z, Junyi G (2013) A tentative study on utilization of canopy hyperspectral reflectance to estimate canopy growth and seed yield in soybean. Acta Agron Sin 39(2):309–318
43. Maes WH, Steppe K (2019) Perspectives for remote sensing with unmanned aerial vehicles in precision agriculture. Trends Plant Sci 24(2):152–164
44. Hui L, Liang T, Wenyu Z, Yanlian W, Weixing C, Yan Z (2009) Construction and implementation of model-based visual rice growth system. Trans Chin Soc Agric Eng 25(9):148–154
45. Yusen Z, Xia Y, Yongchao T, Weixing C, Yan Z (2010) Estimating leaf nitrogen content with near infrared reflectance spectroscopy in rice. Chin J Plant Ecol 34(6):704–712
46. Jun N, Tingting W, Xia Y, Weixing C, Yan Z (2013) Design and experiments of multi-spectral sensor for rice and wheat growth information. Trans Chin Soc Agric Mach 44(5):207–212
47. Zhiyuan H, Yan Z, Yanda L, Xiaojun L, Qiang C, Weixing C, Liang T (2017) Study on estimation model for nitrogen nutrition index and yield on double cropping rice in southern China. J Nanjing Agric Univ 40(1):11–19
48. Xiaojun L, Yongchao T, Xia Y, Weixing C, Yan Z (2012) Monitoring leaf water content based on hyperspectra in rice. Sci Agric Sin 45(3):435–442
49. Yang G, Zhao C, Liu Q, Huang W, Wang J (2011) Inversion of a radiative transfer model for estimating forest LAI from multisource and multi-angular optical remote sensing data. IEEE Trans Geosci Remote Sens 49(3):988–1000
50. Jun N, Xia Y, Yongchao T, Weixing C, Yan Z (2013) Design and experiments of portable apparatus for plant growth monitoring and diagnosis. Trans Chin Soc Agric Eng 29(6):150–156
51. Fu Y, Yang G, Wang J, Song X, Feng H (2014) Winter wheat biomass estimation based on spectral indices, band depth analysis and partial least squares regression using hyperspectral measurements. Comput Electron Agric 100(1):51–59
52. Li P, Wang Q (2011) Retrieval of leaf biochemical parameters using PROSPECT inversion: a new approach for alleviating ill-posed problems. IEEE Trans Geosci Remote Sens 49(7):2499–2506
53. Yanda L, Shifu S, Licai C, Huaping S, Xing L, Song P, Yu L (2013) Photosynthetic production simulation based on plant type in rice: a review. Chin Agric Sci Bull 29(33):1–5
54. Raymond Hunt E Jr, Doraiswamy PC, McMurtrey JE, Daughtry CST, Perry EM, Akhmedov B (2013) A visible band index for remote sensing leaf chlorophyll content at the canopy scale. Int J Appl Earth Obs Geoinf 21(4):103–112
55. Liang T, Xiangcheng Z, Mengying C, Weixing C, Yan Z (2012) Relationships of rice canopy PAR interception and light use efficiency to grain yield. Chin J Appl Ecol 23(5):1269–1276
56. Yebra M, Chuvieco E (2009) Linking ecological information and radiative transfer models to estimate fuel moisture content in the Mediterranean region of Spain: solving the ill-posed inverse problem. Remote Sens Environ 113(11):2403–2411

57. Nguy-Robertson AL, Peng Y, Gitelson AA, Arkebauer TJ, Pimstein A, Herrmann I, Karnieli A, Rundquist DC, Bonfil DJ (2014) Estimating green LAI in four crops: potential of determining optimal spectral bands for a universal algorithm. Agric For Meteorol 192–193(7):140–148
58. Yanda L, Xiangcheng Z, Liang T, Weixing C, Yan Z (2011) Simulation of canopy photosynthetic production based on plant type in rice. Acta Agron Sin 37(5):868–875
59. Yanda L, Liang T, Yuping Z, Leilei L, Weixing C, Yan Z (2010) Spatiotemporal distribution of photosynthetically active radiation in rice canopy. Chin J Appl Ecol 21(4):952–958
60. Weiguo L, Jihua W, Chunjiang Z, Liangyun L, Xiaoyu S, Qingxi T (2008) A model for predicting protein content in winter wheat grain based on Land-Sat TM image and nitrogen accumulation. J Remote Sens 12(3):506–514
61. Liang T, Yanda L, Yuping Z, Xiangcheng Z, Xiaojun L, Weixing C, Yan Z (2011) Simulation of canopy light distribution and application in rice. Chin J Rice Sci 25(4):427–434
62. Zhao C, Wang J, Huang W, Zhou Q (2010) Spectral indices sensitively discriminating wheat genotypes of different canopy architectures. Precis Agric 11(5):557–567
63. Yongchao T, Jie Y, Xia Y, Yan Z, Weixing C (2009) Quantitative relationship between hyperspectral red edge position and canopy leaf nitrogen concentration in rice. Acta Agron Sin 35(9):1681–1690
64. He B, Quan X, Xing M (2013) Retrieval of leaf area index in alpine wetlands using a two-layer canopy reflectance model. Int J Appl Earth Obs Geoinf 21(4):78–91
65. Weiguo L, Hua L, Jihua W, Wenjiang H (2010) A study on classification and monitoring of winter wheat growth status by Landsat/TM image. J Triticeae Crops 30(1):92–95
66. Yanda L, Liang T, Yuping Z, Xiangcheng Z, Weixing C, Yan Z (2010) Relationship of PAR interception of canopy to leaf area and yield in rice. Sci Agric Sin 43(16):3296–3305
67. Shibayama M, Akiyama TA (1986) A spectroradiometer for field use: VI. Radiometric estimation for chlorophyll index of rice canopy. Jpn J Crop Sci 55(4):433–438
68. Liangyun L (2002) Hyperspectral remote sensing application in precision agriculture. Post-Doctoral Dissertation of the Chinese Academy of Sciences, Beijing
69. Cho MA, Skidmore AK (2006) A new technique for extracting the red edge position from hyperspectral data: the linear extrapolation method. Remote Sens Environ 101(2):181–193
70. Chen P, Haboudane D, Tremblay N, Wang J, Vigneault P, Li B (2010) New spectral indicator assessing the efficiency of crop nitrogen treatment in corn and wheat. Remote Sens Environ 114(9):1987–1997
71. Pengfei C, Jiulin S, Jihua W, Chunjiang Z (2010) Using remote sensing technology for crop nitrogen diagnosis: status and tends. Scientia Sinica Informationis 40(Suppl):21–37
72. Chunhua X, Shaokun L, Keru W, Yanli L, Junhua B, Ruizhi X, Shiju G, Xiaojun L, Haizhen T (2008) Response of canopy direction reflectance spectrum for the wheat vertical leaf distribution. Sci Agric Sin 41(8):2271–2278
73. Heli L, Chunjiang Z, Wenjiang H, Yang G (2013) Non-uniform vertical nitrogen distribution within plant canopy and its estimation by remote sensing: a review. Field Crop Res 142(2):75–84
74. Jihua W, Wenjiang H, Cailian L, Luda Z, Changbing L, Tao W, Liangyun L, Xiaoyu S, Zhihong M (2007) Inversion of winter wheat foliage vertical distribution based on canopy reflected spectrum by partial least squares regression method. Spectrosc Spectr Anal 27(7):1319–1322
75. Jihua W, Zhijie W, Wenjiang H, Zhihong M, Liangyun L, Chunjiang Z (2004) The vertical distribution characteristic and spectral response of canopy nitrogen in different layer of winter wheat. J Remote Sens 8(4):309–316
76. Dongyan Z, Xiu W, Craig C, Zhijie W, Jinling Z, Qiang H (2013) Design and experiment of ground-based agriculture-oriented multi-angle observation device. Trans Chin Soc Agric Mach 44(1):174–178
77. Jifeng M, Yan Z, Xia Y, Yongchao T, Xiaojun L, Weixing C (2007) Relationship between leaf nitrogen content and fluorescence parameters in rice. Chin J Rice Sci 21(1):65–70
78. Wu C, Han X, Niu Z, Dong J (2010) An evaluation of EO-1 hyperspectral Hyperion data for chlorophyll content and leaf area index estimation. Int J Remote Sens 31(4):1079–1086

79. Xin Q, Xu M, Xiaochao Z, Liusuo L (2008) Response of coronary spectrum on chlorophyll and K information of soy. Trans Chin Soc Agric Mach 39(4):108–111
80. Wu C, Wang L, Niu Z, Gao S, Wu M (2010) Nondestructive estimation of canopy chlorophyll content using Hyperion and Landsat/TM images. Int J Remote Sens 31(8):2159–2167
81. Kitchen NR, Sudduth KA, Drummond ST, Scharf PC, Palm HL, Roberts DF, Vories ED (2010) Ground-based canopy reflectance sensing for variable-rate nitrogen corn fertilization. Agron J 102(1):71–84
82. Peng S, Garcia FV, Laza RC, Sanico AL, Visperas RM, Cassman KG (1996) Increased N-use efficiency using a chlorophyll meter on high-yielding irrigated rice. Field Crops Res 47(2–3):243–252
83. Scharf PC, Lory JA (2002) Calibrating corn color from aerial photographs to predict sidedress nitrogen need. Agron J 94(3):397–404
84. Lukina EV, Freeman KW, Wynn KJ, Thomason WE, Mullen RW, Stone ML, Solie JB, Klatt AR, Johnson GV, Elliott RL, Raun WR (2001) Nitrogen fertilization optimization algorithm based on in-season estimates of yield and plant nitrogen uptake. J Plant Nutr 24(6):885–898
85. Liping C (2003) Theoretical and experimental studies on variable-rate fertilization in precision farming. China Agricultural University, Beijing
86. A'ning J, Wenjiang H, Chunjiang Z, Keli L, Liangyun L, Jihua W (2007) Effects of variable nitrogen application based on characteristics of canopy light reflectance in wheat. Sci Agric Sin 40(9):1907–1913
87. Ji Z, Tardieu F, Pridmore T, Doonan J, Reynolds D, Hall N, Griffiths S, Tao C, Yan Z, Xiu'e W, Dong J, Yanfeng D (2018) Plant phenomics: history, present status and challenges. J Nanjing Agric Univ 41(4):580–588
88. Jinglu W, Ying Z, Xiaodi P, Xianju L, Liming M, Xinyu G (2018) Research progress and prospect on crop phenomics database. China Agric Inf 30(5):13–23
89. Qiu R, Wei S, Zhang M, Li H, Sun H, Liu G, Li M (2018) Sensors for measuring plant phenotyping: a review. Int J Agric Biol Eng 11(2):1–17
90. Jiangang L, Chunjiang Z, Guijun Y, Haiyang Y, Xiaoqing Z, Bo X, Qinglin N (2016) Review of field-based phenotyping by unmanned aerial vehicle remote sensing platform. Trans Chin Soc Agric Eng 32(24):98–106
91. Córcoles JI, Ortega JF, Hernández D, Moreno MA (2013) Estimation of leaf area index in onion (Allium cepa L.) using an unmanned aerial vehicle. Biosyst Eng 115(1):31–42
92. Duan T, Zheng B, Guo W, Ninomiya S, Guo Y, Chapman SC (2017) Comparison of ground cover estimates from experiment plots in cotton, sorghum and sugarcane based on images and ortho-mosaics captured by UAV. Funct Plant Biol 44(1):169–183
93. Rasmussen J, Ntakos G, Nielsen J, Svensgaard J, Poulsen RN, Christensen S (2016) Are vegetation indices derived from consumer-grade cameras mounted on UAVs sufficiently reliable for assessing experimental plots? Eur J Agron 74(3):75–92
94. Jin X, Liu S, Baret F, Hemerlé M, Comar A (2017) Estimates of plant density of wheat crops at emergence from very low altitude UAV imagery. Remote Sens Environ 198(9):105–114
95. Liu S, Baret F, Andrieu B, Burger P, Hemmerlé M (2017) Estimation of wheat plant density at early stages using high resolution imagery. Front Plant Sci 8(5):739–749
96. Xu R, Li C, Paterson AH, Jiang Y, Sun S, Robertson JS (2018) Aerial images and convolutional neural network for cotton bloom detection. Front Plant Sci 8(2):2235–2252
97. Bucksch A, Burridge J, York LM, Das A, Nord E, Weitz JS, Lynch JP (2014) Image-based high-throughput field phenotyping of crop roots. Plant Physiol 166(10):470–486
98. Zhang J, Yang C, Zhao B, Song H, Hoffmann WC, Shi Y, Zhang D, Zhang G (2017) Crop classification and LAI estimation using original and resolution-reduced images from two consumer-grade cameras. Remote Sens 9(10):1054–1072
99. Raymond Hunt E Jr, Cavigelli M, Daughtry CST, Mcmurtrey JE III, Walthall CL (2005) Evaluation of digital photography from model aircraft for remote sensing of crop biomass and nitrogen status. Precis Agric 6(4):359–378

100. Zhou X, Zheng HB, Xu XQ, He JY, Ge XK, Yao X, Cheng T, Zhu Y, Cao WX, Tian YC (2017) Predicting grain yield in rice using multi-temporal vegetation indices from UAV-based multispectral and digital imagery. ISPRS J Photogramm Remote Sens 130(8):246–255
101. Li L, Zhang Q, Huang D (2014) A review of imaging techniques for plant phenotyping. Sensors 14(11):20078–20111
102. Fahlgren N, Gehan MA, Baxter I (2015) Lights, camera, action: high-throughput plant phenotyping is ready for a close-up. Curr Opin Plant Biol 24(4):93–99
103. Duan T, Chapman SC, Guo Y, Zheng B (2017) Dynamic monitoring of NDVI in wheat agronomy and breeding trials using an unmanned aerial vehicle. Field Crop Res 210(15):71–80
104. Großkinsky DK, Svensgaard J, Christensen S, Roitsch T (2015) Plant phenomics and the need for physiological phenotyping across scales to narrow the genotype-to-phenotype knowledge gap. J Exp Bot 66(18):5429–5440
105. Yingxue L, Yan Z, Yyongchao T, Xia Y, Changjun Z, Weixing C (2006) Quantitative relationship between leaf nitrogen accumulation and canopy reflectance spectra in wheat. Acta Agron Sin 32(2):203–209
106. Read JJ, Tarpley L, McKinion JM, Reddy KR (2002) Narrow-waveband reflectance ratios for remote estimation of nitrogen status in cotton. J Environ Qual 31(5):1442–1452
107. Lihong X, Weixing C, Weihong L, Dong J, Yali M, Yan Z (2003) Diagnosis of nitrogen status in rice leaves with canopy spectral reflectance. Sci Agric Sin 2(3):250–257
108. Petisco C, García-Criado B, Mediavilla S, Vázquez de Aldana BR, Zabalgogeazcoa I, García-Ciudad A (2006) Near-infrared reflectance spectroscopy as a fast and non-destructive tool to predict foliar organic constituents of several woody species. Anal Bioanal Chem 386(6):1823–1833
109. Huanyu J, Yibin Y (2007) Measurement of chlorophyll content in pepper leaves by near infrared analysis. Spectrosc Spectr Anal 27(3):499–502
110. Luda Z, Xiaonan S, Longlian Z, Junhui L, Jianping Z, Wenyan X, Ruxin S (2000) Application of principal component-all possible regression in quantitative analysis of flue-cured tobacco and wheat using near-infrared spectroscopy. Chin J Anal Chem 28(6):723–726
111. Qian C, Jingfeng H, Renchao W, Yanlin T (2004) Correlation analysis of simulated MODIS vegetation indices and rice leaf area index and leaf chlorophyll content. Chin J Appl Ecol 15(8):1363–1367
112. Baluja J, Diago MP, Balda P, Zorer R, Meggio F, Morales F, Tardaguila J (2012) Assessment of vineyard water status variability by thermal and multispectral imagery using an unmanned aerial vehicle (UAV). Irrig Sci 30(6):511–522
113. Yongchao T, Jie Y, Xia Y, WeiXing C, Yan Z (2010) Monitoring canopy leaf nitrogen concentration based on leaf hyperspectral indices in rice. Acta Agron Sin 36(9):1529–1537
114. Weijun C, Qifa Z, Jingfeng H (2006) Estimating pigment contents in leaves and panicles of rice after milky ripening by hyperspectral vegetation indices. Chin J Rice Sci 20(4):434–439
115. Adão T, Hruška J, Bessa LPJ, Peres E, Morais R, Sousa JJ (2017) Hyperspectral imaging: a review on UAV-based sensors, data processing and applications for agriculture and forestry. Remote Sens 9(11):1–30
116. Mahlein A-K, Steiner U, Hillnhütter C, Dehne H-W, Oerke E-C (2012) Hyperspectral imaging for small-scale analysis of symptoms caused by different sugar beet diseases. Plant Methods 24(8):3–15
117. Li J, Rao X, Ying Y (2011) Detection of common defects on oranges using hyperspectral reflectance imaging. Comput Electron Agric 78(1):38–48
118. Mahlein A-K (2016) Plant disease detection by imaging sensors-parallels and specific demands for precision agriculture and plant phenotyping. Irrig Sci 100(2):241–251
119. Fiorani F, Rascher U, Jahnke S, Schurr U (2012) Imaging plants dynamics in heterogenic environments. Curr Opin Biotechnol 23(2):227–235
120. Hui F, Lizhong X, Guoxing C, Wanneng Y, Qian L (2015) The segmentation of leaf and stem of individual rice plant with hyperspectral imaging system and principal component analysis. Acta Laser Biol Sin 24(1):31–38

121. Mei-ling L, Fei D, Ying L, Hengnian Q, Xiao Z (2015) Study on detection technology of rice seed vigor based on hyperspectral image. Acta Agriculturae Zhejiangensis 27(1):1–6
122. Huang W, Lamb DW, Niu Z, Zhang Y, Liu L, Wang J (2007) Identification of yellow rust in wheat using in-situ spectral reflectance measurements and airborne hyperspectral imaging. Precis Agric 8(4–5):187–197
123. Nansen C, Zhao G, Dakin N, Zhao C, Turner SR (2015) Using hyperspectral imaging to determine germination of native Australian plant seeds. J Photochem Photobiol B Biol 145(4):19–24
124. Xiaolong X, Huanyu J, Yuelan H (2012) Study on detection of tomato mosaic disease at early stage based on infrared thermal imaging. Trans Chin Soc Agric Eng 28(5):145–149
125. Cohen Y, Alchanatis V, Meron M, Saranga Y, Tsipris J (2005) Estimation of leaf water potential by thermal imagery and spatial analysis. J Exp Bot 56(417):1843–1852
126. Jones HG, Serraj R, Loveys BR, Xiong L, Wheaton A, Price AH (2009) Thermal infrared imaging of crop canopies for the remote diagnosis and quantification of plant responses to water stress in the field. Funct Plant Biol 36(11):978–989
127. Berni JAJ, Zarco-Tejada PJ, Suarez L, Fereres E (2009) Thermal and narrowband multispectral remote sensing for vegetation monitoring from an unmanned aerial vehicle. IEEE Trans Geosci Remote Sens 47(3):722–738
128. Siebert S, Ewert F, Rezaei EE, Kage H, Graß R (2014) Impact of heat stress on crop yield-on the importance of considering canopy temperature. Environ Res Lett 9(4):1–8
129. Gonzalez-Dugo V, Hernandez P, Solis I, Zarco-Tejada PJ (2015) Using high-resolution hyperspectral and thermal airborne imagery to assess physiological condition in the context of wheat phenotyping. Remote Sens 7(10):13586–13605
130. Lelong CCD, Burger P, Jubelin G, Roux B, Labbé S, Baret F (2008) Assessment of unmanned aerial vehicles imagery for quantitative monitoring of wheat crop in small plots. Sensors 8(5):3557–3585
131. Costa JM, Grant OM, Chaves MM (2013) Thermography to explore plant–environment interactions. J Exp Bot 64(13):3937–3949
132. Berni JAJ, Zarco-Tejada PJ, Sepulcre-Cantó G, Fereres E, Villalobos F (2009) Mapping canopy conductance and CWSI in olive orchards using high resolution thermal remote sensing imagery. Remote Sens Environ 113(11):2380–2388
133. Sullivan DG, Fulton JP, Shaw JN, Bland G (2007) Evaluating the sensitivity of an unmanned thermal infrared aerial system to detect water stress in a cotton canopy. Trans ASABE 50(6):1963–1969
134. Awad YM, Abdullah AA, Bayoumi TY, Abd-Elsalam K, Hassanien AE (2014) Early detection of powdery mildew disease in wheat (Triticum aestivum L.) using thermal imaging technique. Int Underw Syst Des 323(1):755–765
135. Maes WH, Huete AR, Steppe K (2017) Optimizing the processing of UAV-based thermal imagery. Remote Sens 9(5):476–493
136. Prashar A, Jones HG (2014) Infra-red thermography as a high-throughput tool for field phenotyping. Agronomy 4(3):397–417
137. Ya L (2012) Application of infrared thermography in the research of plant mechanism response to drought. Chin Agric Sci Bull 28(3):17–22
138. Daley PF (1995) Chlorophyll fluorescence analysis and imaging in plant stress and disease. Can J Plant Pathol 17(2):167–173
139. Jinzhu L, Huanyu J, Di C (2014) Progress of fluorescence imaging technology in detection of plant diseases. Trans Chin Soc Agric Mach 45(4):244–252
140. Chunping W, Kairong L, Zhengguo L, Qing L, Hong W (2012) Effects of chilling stress on chlorophyll fluorescence characteristics of seedling leaves with different leaf ages of Oryza sativa. J Plant Resour Environ 21(3):38–43
141. Harbinson J, Prinzenberg AE, Kruijer W, Aarts MGM (2012) High throughput screening with chlorophyll fluorescence imaging and its use in crop improvement. Curr Opin Biotechnol 23(2):221–226

142. Lee JH, Lee JH, Jung SH, Hyun TK, Feng M, Kim J-Y, Lee J-H, Lee H, Kim JS, Kang C, Kwon K-Y, Jung JH (2015) Highly selective fluorescence imaging of zinc distribution in HeLa cells and Arabidopsis using a naphthalene-based fluorescent probe, vol 51(3). The Royal Society of Chemistry, London, pp 7463–7465
143. Bauriegel E, Giebel A, Geyer M, Schmidt U, Herppich WB (2011) Early detection of Fusarium infection in wheat using hyper-spectral imaging. Comput Electron Agric 75(2):304–312
144. Zheng B, Shi L, Ma Y, Deng Q, Li B, Guo Y (2008) Comparison of architecture among different cultivars of hybrid rice using a spatial light model based on 3-D digitising. Funct Plant Biol 35(10):900–910
145. Bangyou Z, Lijuan S, Yuntao M, Qiyun D, Baoguo L, Yan G (2009) Three-dimensional digitization in situ of rice canopies and virtual stratified-clipping method. Sci Agric Sin 42(4):1181–1189
146. Alenya G, Dellen B, Foix S, Torras C (2013) Robotized plant probing: leaf segmentation utilizing time-of-flight data. IEEE Robot Autom Mag 20(3):50–59
147. Deery D, Jimenez-Berni J, Jones H, Sirault X, Furbank R (2014) Proximal remote sensing buggies and potential applications for field-based phenotyping. Agronomy 4(3):349–379
148. Greaves HE, Vierling LA, Eitel JUH, Boelman NT, Magney TS, Prager CM, Griffin KL (2015) Estimating aboveground biomass and leaf area of low-stature arctic shrubs with terrestrial LiDAR. Remote Sens Environ 164(7):26–35
149. Jay S, Rabatel G, Hadoux X, Moura D, Gorretta N (2015) In-field crop row phenotyping from 3D modeling performed using structure from motion. Comput Electron Agric 110(1):70–77
150. Pengcheng H, Yan G, Baoguo L, Jinyu Z, Yuntao M (2015) Three-dimensional reconstruction and its precision evaluation of plant architecture based on multiple view stereo method. Trans Chin Soc Agric Eng 31(11):209–214
151. Jiang Y, Li C, Paterson AH (2016) High throughput phenotyping of cotton plant height using depth images under field conditions. Comput Electron Agric 130(15):57–68
152. Qinghua G, Fangfang W, Shuxin P, Xiaoqian Z, Linhai C, Jin L, Baolin X, Guangcai X, Le L, HaiChun J, Chengcai C (2016) Crop 3D: a platform based on LiDAR for 3D high-throughput crop phenotyping. Scientia Sinica Vitae 46(10):1210–1221
153. Honkavaara E, Saari H, Kaivosoja J, Pölönen I, Hakala T, Litkey P, Mäkynen J, Pesonen L (2013) Processing and assessment of spectrometric, stereoscopic imagery collected using a lightweight UAV spectral camera for precision agriculture. Remote Sens 5(10):5006–5039
154. Moecke T, Safari H, Reddersen B, Fricke T, Wachendorf M (2017) Fusion of ultrasonic and spectral sensor data for improving the estimation of biomass in grasslands with heterogeneous sward structure. Remote Sens 9(1):98–112
155. Zhao K, Popescu S, Nelson R (2009) Lidar remote sensing of forest biomass: a scale-invariant estimation approach using airborne lasers. Remote Sens Environ 113(1):182–196
156. Kipp S, Mistele B, Schmidhalter U (2014) The performance of active spectral reflectance sensors as influenced by measuring distance, device temperature and light intensity. Comput Electron Agric 100(2):24–33
157. Jian Z, Zongnan L, Nan Z, Jing X, Liyuan H (2013) Advances in 3D information collection and reconstruction of crop based on the measured data. J Huazhong Agric Univ 32(4):126–134
158. Xihong C, Jin C, Linlin G (2009) The application of ground penetrating radar to plant root system detection. Adv Earth Science 24(6):606–611
159. Li G, Bihang F, Yuan W, Wentao L, Xihong C, Jin C (2014) A review on the application of ground-penetrating radar to detect and quantify coarse roots. Sciencepaper Online 9(4):494–498
160. Rui L, Ting L, Runru D, Zhi L, Dehai Z, Wei S (2014) 3D modeling of maize based on terrestrial LiDAR point cloud data. J China Agric Univ 19(3):196–201
161. Qian S, Qingbo Z, Wenbin W, Qiong H, Qiangyi Y, Huajun T (2015) Recent progresses in research of integrating multi-source remote sensing data for crop mapping. Sci Agric Sin 48(6):1122–1135

162. Qinghua G, Weicai Y, Fangfang W, Shuxin P, Shichao J, Fan C, Xiujie W (2018) High-throughput crop phenotyping: accelerators for development of breeding and precision agriculture. Bull Chin Acad Sci 33(9):940–946
163. Sankaran S, Khot LR, Espinoza CZ, Jarolmasjed S, Sathuvalli VR, Vandemark GJ, Miklas PN, Carter AH, Pumphrey MO, Richard Knowles N, Pavek MJ (2015) Low-altitude, high-resolution aerial imaging systems for row and field crop phenotyping: a review. Eur J Agron 70(8):112–123
164. Boegh E, Soegaard H, Broge N, Hasager CB, Jensen NO, Schelde K, Thomsen A (2002) Airborne multispectral data for quantifying leaf area index, nitrogen concentration, and photosynthetic efficiency in agriculture. Remote Sens Environ 81(2–3):179–193
165. Deery DM, Rebetzke GJ, Jimenez-Berni JA, James RA, Condon AG, Bovill WD, Hutchinson P, Scarrow J, Davy R, Furbank RT (2016) Methodology for high-throughput field phenotyping of canopy temperature using airborne thermography. Front Plant Sci 7(12):1808–1821
166. Rajan N, Maas SJ (2009) Mapping crop ground cover using airborne multispectral digital imagery. Precis Agric 10(4):304–318
167. Chapman SC, Merz T, Chan A, Jackway P, Hrabar S, Fernanda Dreccer M, Holland E, Zheng B, Jun Ling T, Jimenez-Berni J (2014) Pheno-copter: a low-altitude, autonomous remote-sensing robotic helicopter for high-throughput field-based phenotyping. Agronomy 4(2):279–301
168. Jonckheere I, Fleck S, Nackaerts K, Muys B, Coppin P, Weiss M, Baret F (2004) Review of methods for in situ leaf area index determination: part I. Theories, sensors and hemispherical photography. Agric For Meteorol 121(1–2):19–35
169. Krienke B, Ferguson RB, Schlemmer M, Holland K, Marx D, Eskridge K (2017) Using an unmanned aerial vehicle to evaluate nitrogen variability and height effect with an active crop canopy sensor. Precis Agric 18(6):900–915
170. Changhua L, Zhe W, Zhichao C, Lan Z, Xuezhi Y, Yuxin M (2018) Nitrogen monitoring of winter wheat based on unmanned aerial vehicle remote sensing image. Trans Chin Soc Agric Mach 49(6):207–214
171. Zheng H, Cheng T, Li D, Yao X, Tian Y, Cao W, Zhu Y (2018) Combining unmanned aerial vehicle (UAV)-based multispectral imagery and ground-based hyperspectral data for plant nitrogen concentration estimation in rice. Front Plant Sci 9(7):936–949
172. Konishi N, Shiga H, Asaka D et al (2000) The application of remote sensing technology in high quality rice production. Measurement 10(7):14–20. (In Japanese)
173. Araus JL, Cairns JE (2014) Field high-throughput phenotyping: the new crop breeding frontier. Trends Plant Sci 19(1):52–61
174. Duan T, Chapman SC, Holland E, Rebetzke GJ, Guo Y, Zheng B (2016) Dynamic quantification of canopy structure to characterize early plant vigour in wheat genotypes. J Exp Bot 67(15):4523–4534
175. Pengcheng H, Chapman SC, Wang X, Potgieter A, Duan T, Jordan D, Guo Y, Zheng B (2018) Estimation of plant height using a high throughput phenotyping platform based on unmanned aerial vehicle and self-calibration: example for sorghum breeding. Eur J Agron 95(4):24–32
176. Busemeyer L, Mentrup D, Möller K, Wunder E, Alheit K, Hahn V, Maurer HP, Reif JC, Würschum T, Müller J, Rahe F, Ruckelshausen A (2013) BreedVision—a multi-sensor platform for non-destructive field-based phenotyping in plant breeding. Sensors 13(3):2830–2847
177. Sharma LK, Bu H, Franzen DW, Denton A (2016) Use of corn height measured with an acoustic sensor improves yield estimation with ground based active optical sensors. Comput Electron Agric 124(6):254–262
178. Minglu T, Songtao B, Qingrui C, Mingming Y, Dan L, Li W, Shuo W (2016) Use of hyperspectral images from UAV-based imaging spectroradiometer to estimate cotton leaf area index. Trans Chin Soc Agric Eng 32(21):102–108
179. Rebetzke GJ, Jimenez-Berni JA, Bovill WD, Deery DM, James RA (2016) High-throughput phenotyping technologies allow accurate selection of stay-green. J Exp Bot 67(17):4919–4924

180. Sharma B, Ritchie GL (2015) High-throughput phenotyping of cotton in multiple irrigation environments. Crop Sci 55(2):958–969
181. Jie J, Zeyu Z, Qiang C, Yongchao T, Yan Z, Weixing C, Xiaojun L (2019) Use of a digital camera mounted on a consumer-grade unmanned aerial vehicle to monitor the growth status of wheat. J Nanjing Agric Univ 42(4):622–631
182. Dammer K-H, Wollny J, Giebel A (2008) Estimation of the leaf area index in cereal crops for variable rate fungicide spraying. Eur J Agron 28(3):351–360
183. Ehlert D, Adamek R, Horn H-J (2009) Laser rangefinder-based measuring of crop biomass under field conditions. Precis Agric 10(5):395–408
184. Ramirez MB (2010) Monitoring nitrogen levels in the cotton canopy using real-time active-illumination spectral sensing. University of Tennessee-Knoxville, Knoxville
185. Lee W, Searcy SW. Multispectral sensor for detecting nitrogen in corn plants. In: 2000 ASAE annual international meeting sponsored by ASAE, July 9-12, 2000
186. Debuisson S, Germain C, Garcia O, Panigai L, Moncomble D. Using multiplex® and greens-eekerTM to manage spatial variation of vine vigor in champagne
187. Sudduth KA, Kitchen NR, Drummond ST. Nadir and oblique canopy reflectance sensing for N application in corn. In: 2011 ASABE annual international meeting sponsored by ASABE, August 7-10, 2011
188. Yanwei Y, Shujun L, Xianfa F, Liguo W, Yangchun L (2013) Decision support system of N, P and K ratio fertilization. Trans Chin Soc Agric Mach 44(8):240–244, 223
189. Xin Q (2008) Research of soybean nutrition information diagnosis based on spectrum technology and system of variable rate fertilizer. Jilin University, Changchun
190. Rui Z (2012) The research on key technology of intelligence variable rate fertilization. Chinese Academy of Agricultural Sciences, Beijing

Chapter 2
Precision Variable-Rate Fertilization for Rice-Wheat Cropping Using Outer-Grooved Wheel Mechanism Based on Multi-layer Perceptron Model

Xuekai Huang, Yinyan Shi, D. Wang, J. Bo, Man Chen, and L. Lin

Abstract To address the limited focus on actuator mechanisms in current variable rate fertilization (VRF) research, this study presents a novel segmented-shaft actuator for outer-grooved wheel fertilizer applicators, enabling dual-variable control of rotor speed and opening size for real-time adjustment of fertilizer dosage. A discrete element model was developed using EDEM 2.2 to simulate fertilizer discharge behavior under different structural configurations and control strategies. Simulation results were validated through bench and field experiments, demonstrating that the improved device reduced discharge variability by 9.9%, with the optimal variation coefficient reaching 1.85% under speed-priority control. To enhance control precision and real-time responsiveness, a hybrid predictive model based on a multilayer perceptron (MLP) neural network was constructed and optimized using the Levy Flight Algorithm (LFA) and Particle Swarm Optimization (PSO), forming the LFA-PSO-MLP (LMP) model. Additionally, an inverse LMP (ILMP) model was established for closed-loop computation of actuator parameters from target fertilizer rates. The LMP model achieved rapid convergence (≈50 iterations), high fitting accuracy ($R^2 = 0.999$), and low mean absolute percentage error (1.83%). Field trials further confirmed a mean application accuracy of 93.92%. These results demon-

X. Huang · D. Wang · J. Bo · L. Lin
College of Engineering, Nanjing Agricultural University, Nanjing, China
e-mail: 2022212011@stu.njau.edu.cn; 2024212004@stu.njau.edu.cn; 2022212013@stu.njau.cn; 2023212009@stu.njau.edu.cn

Y. Shi (✉)
College of Engineering, Nanjing Agricultural University, Nanjing, China

Jiangsu Province Engineering Lab for Modern Facility Agriculture Technology & Equipment, Nanjing, China
e-mail: shiyinyan@njau.edu.cn

M. Chen
Nanjing Research Institute for Agricultural Mechanization, Ministry of Agriculture and Rural Affairs, Nanjing, China
e-mail: chenman@caas.cn

Y. Shi et al. (eds.), *Technological Progress on Variable Rate Fertilization*, Smart Agriculture 17, https://doi.org/10.1007/978-981-95-4330-4_2

strate the feasibility and effectiveness of the proposed actuator design and intelligent modeling approach for improving precision fertilization in rice-wheat cropping systems, contributing to sustainable and intelligent agricultural machinery development.

Keywords Variable rate fertilization · Outer-grooved wheel · Multilayer perceptron · Discrete element modeling · Actuator design · Neural network prediction

2.1 Introduction

Precision agriculture provides a technological foundation for sustainable agricultural development. Among its key components, variable rate fertilization (VRF) integrates 3S technologies (GPS, GIS, RS) with agronomic and soil science knowledge to optimize nutrient exchange within the crop-soil system, thereby improving fertilizer utilization, reducing production costs, and minimizing environmental pollution [1–3]. In recent years, VRF techniques have witnessed substantial development globally, with increasing adoption of intelligent data acquisition, processing, and decision-making systems. However, research on the core mechanical components—specifically, the actuators responsible for fertilizer discharge—remains limited.

As the critical component of VRF equipment, the actuator directly determines the precision and stability of fertilizer application. Existing actuators mainly include electro-mechanical variable transmission systems, electro-hydraulic drive systems, and direct motor-driven mechanisms. Among these, stepper motor-driven actuators offer structural simplicity, lower cost, torque reliability, and robust field adaptability [4, 5]. Accordingly, this study develops a novel segmented-shaft dual-regulation actuator, adopting a stepper motor for direct control of both the rotor speed and the fertilizer outlet opening.

Traditional fertilizer applicators often suffer from complex operations, low stability, and delayed responsiveness, which hinder widespread adoption. The proposed actuator addresses these shortcomings by supporting automation, simplifying operations, and meeting the topdressing requirements of winter wheat and other staple crops. In rice–wheat rotation regions of the Yangtze River basin, outer-grooved wheel applicators remain widely used [6, 7]. However, due to their pulsation at low rates and speeds, they often introduce intermittent discharge, severely affecting precision [7].

To address these limitations, discrete element modeling (DEM) has gained traction for analyzing particle-soil-equipment interactions. Previous studies applied DEM to optimize fertilizer mixing, discharge uniformity, and force interaction mechanisms in grain application [7–9]. Building on these advances, this study develops a DEM simulation model for the segmented outer-grooved actuator using EDEM software to evaluate the influence of structural variations and control strategies on discharge stability. Bench and field experiments are conducted to validate simulation accuracy and improve application performance [10–12].

In parallel, establishing a predictive model for fertilizer discharge that captures the nonlinear relationship between outlet opening, rotor speed, and discharge rate is essential for achieving high-efficiency variable fertilization [13, 14]. While statistical models, machine learning, deep learning, and GIS-based spatial prediction have been explored, machine learning models strike a balance between accuracy, data requirements, and computational efficiency [15–17]. Therefore, this study employs a multilayer perceptron (MLP) neural network model optimized by Levy Flight Algorithm (LFA) and Particle Swarm Optimization (PSO) to build a forward discharge prediction model (LFA-PSO-MLP, LMP) and its inverse control model (ILMP) [18–22].

Existing research largely focuses on single-variable prediction for crop nutrient demand, with limited attention to dual-variable actuator control for discharge prediction in rice–wheat systems. This study bridges that gap by constructing and validating a fast, accurate, and robust discharge prediction framework based on dual-input parameters (opening and speed) to meet the demands of modern precision agriculture.

2.2 Overall Structure and Working Principle

2.2.1 Overall Structure

The novel segmented-shaft dual-regulation variable-rate fertilizer applicator designed in this study is shown in Fig. 2.1. It adopts a suspended traction configuration with a working width of 2 m, consisting of two independent working groups, each covering a 1-m operation zone. The input shaft that drives the rotation of the outer-grooved wheel fertilizer applicators is directly connected to a brushless DC planetary gear motor via a coupling. A slider on a leadscrew rail, driven by a 57-series stepper motor, is mechanically linked to the input shaft through a lever mechanism.

The fertilizer controller receives real-time data from a hyperspectral sensor and, in combination with an expert fertilization system, determines the required fertilizer application rate [23, 24]. Considering the forward speed of the implement, the system converts the desired rate into a combination of wheel rotation speed and fertilizer outlet opening for the outer-grooved wheel applicators. As shown in the 3D view of the variable-rate applicator, the eight fertilizer discharge units are arranged into left and right groups of four and mounted on separate segmented input shafts. These shafts extend horizontally through each fertilizer unit and are internally fitted with outer-grooved wheels. One end of each input shaft is coupled to the output shaft of a brushless DC planetary gear motor, which is bolted onto a fixed motor mount and drives the rotation of the input shaft to control the applicator's rotation speed [25, 26].

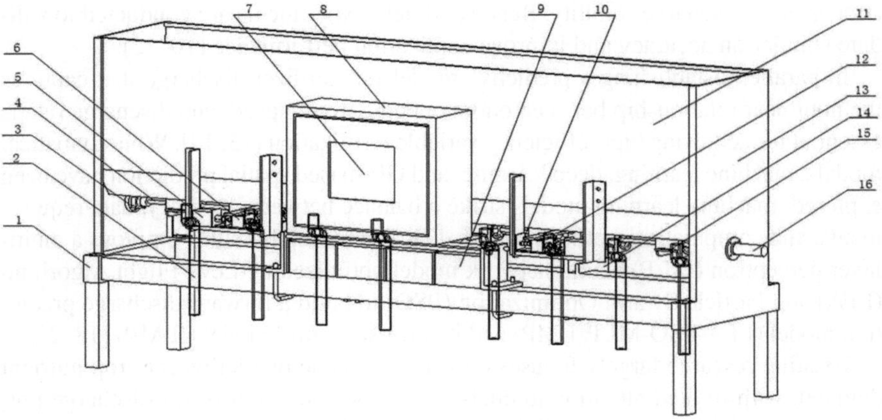

Fig. 2.1 Structure diagram of variable rate fertilizer machinery. 1. Bracket, 2. Fertilizer hose, 3. Coupling, 4. Main shaft, 5. Stepper motor, 6. Slider, 7. Display screen, 8. Electrical control box, 9. Battery, 10. Leadscrew, 11. Fertilizer tank, 12. Lever, 13. Baffle, 14. Suspension, 15. Grooved wheel, 16. Gear reduction motor

A stepper motor controls a leadscrew slider mounted on a guide rail. This slider is coupled to the input shaft of the fertilizer applicator via a connecting sleeve. As the stepper motor rotates, the slider moves along the axis, and this movement is transferred through a lever to a main shaft slider, enabling axial movement of the grooved wheel. This mechanism adjusts the effective working length of the outer-grooved wheel, thereby regulating the fertilizer discharge volume. The lever mechanism converts the rotational motion of the stepper motor into linear motion of the grooved wheel along the shaft, with the range of movement constrained by limit switches installed on the leadscrew (Fig. 2.2).

2.2.2 *Working Principle of the Fertilizer Applicator Actuator*

The segmented-shaft dual-regulation fertilizer applicator actuator developed in this study consists of a stepper motor, leadscrew rail, lever, and coupling, and adopts a symmetrical dual-input shaft configuration. The stepper motor drives the leadscrew slider, which, through the lever mechanism, induces lateral displacement of the input shaft, thereby adjusting the effective working length of the outer-grooved wheel fertilizer applicator. A displacement sensor monitors the slider's position in real time, while limit switches at both ends of the leadscrew ensure safe travel limits.

1. Sensor 2.Onboard CPU 3.DC gear motor 4. Stepper motor 5.Battery 6.Control box

Fig. 2.2 Schematic diagram of variable rate fertilizer applicator. 1. Sensor, 2. Onboard CPU, 3. DC gear motor, 4. Stepper motor, 5. Battery, 6. Control box

For discharge rate control, the input shaft is directly coupled to a brushless DC planetary gear motor. The controller generates PWM signals to regulate the motor's speed, enabling precise adjustment of the fertilizer application rate.

During field operations, a proximal hyperspectral sensor captures NDVI data from the crop canopy and transmits it to the onboard CPU. Combined with real-time speed and positioning data, an embedded Raun fertilization model computes the target application rate. This target value is input into the ILMP neural network model, which calculates the optimal combination of actuator opening and rotation speed. The resulting control commands are sent to the controller, which adjusts the actuator accordingly. Through continuous sensing and feedback, the system achieves closed-loop control for high-precision variable-rate fertilization in rice–wheat cropping systems (Fig. 2.3).

2.2.3 Structure Design and Improvement of the Fertilizer Dispenser

The structural design of the fertilizer dispenser plays a critical role in determining the accuracy and consistency of variable-rate fertilization. Based on operational issues encountered during previous field applications—particularly fertilizer flow instability caused by groove wheel dispensers—this study investigates the structure of a typical outer-grooved wheel dispenser, as shown in Fig. 2.4.

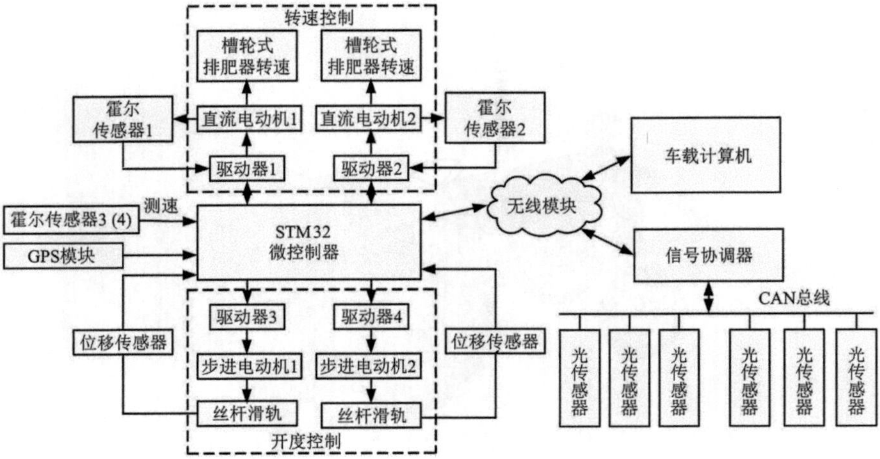

Fig. 2.3 Schematic diagram of variable fertilization principle

Fig. 2.4 Structure diagram of typical outer groove-wheel

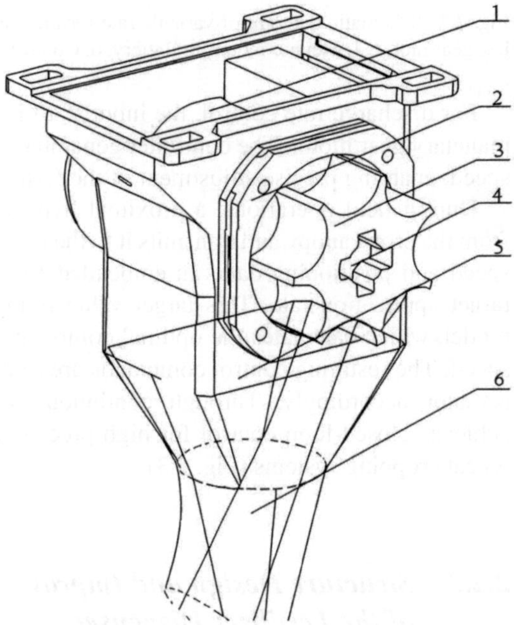

To address the problems of high discharge pulsation and poor operational stability, structural modifications were implemented on the basis of experimental analyses of the conventional design. The improvement process is illustrated in Fig. 2.5.

As depicted in Fig. 2.5a, the original design adopts a traditional seven-groove outer-grooved wheel with an effective groove length of 45.8 mm and a wheel diameter of 58 mm. The dispenser shaft has a diameter of 17 mm, the fertilizer chamber width is 45.16 mm, and the outlet diameter is 41 mm. The working groove wheel (1)

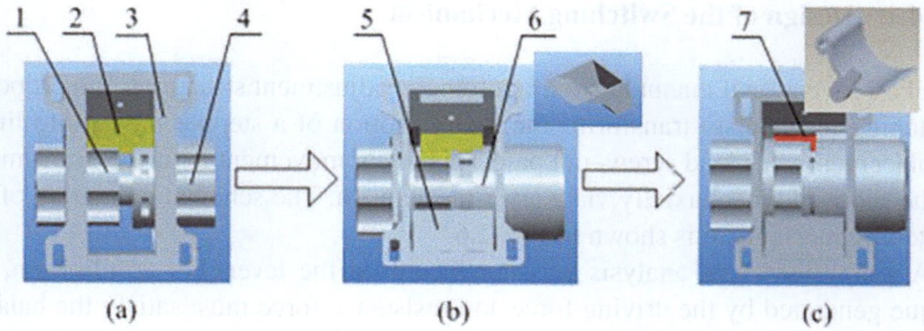

Fig. 2.5 Structure comparison diagram of outer groove-wheel. (**a**) The original design of the grooved-wheel fertilizer applicator. (**b**) The improved design with an integrally machined groove wheel. (**c**) The further optimized design featuring a fertilizer tongue

and the adjusting groove wheel (4) are separate components. Fertilizer discharge is regulated by a plastic brush in contact with the groove wheel. However, during high-speed operations, excessive friction leads to elevated temperatures, resulting in thermal deformation and enlarged gaps between the components. This condition often causes fertilizer leakage and significant variation in the discharge rate, adversely affecting application stability.

In the improved design (Fig. 2.5b), the groove wheels are integrally machined, and additional components such as a blocking sleeve and discharge baffle are incorporated. The inner diameter of the blocking sleeve is 52 mm, while the outer diameter matches that of the groove wheel (58 mm). The baffle width is aligned with the fertilizer chamber (45.16 mm), effectively constraining fertilizer flow within the desired channel.

A further improvement is shown in Fig. 2.5c, where the plastic brush is replaced with a fertilizer tongue, 45 mm in width. This tongue, aided by spring tension, maintains reliable sealing and chamber filling, thereby reducing leakage and ensuring smoother fertilizer delivery. These structural enhancements significantly improve the discharge performance and operational reliability of the fertilizer dispenser under variable-rate conditions.

2.2.4 Design of the Actuator for the Novel Segmented-Shaft Dual-Regulation Variable-Rate Fertilizer Applicator

The actuator of the proposed segmented-shaft variable-rate fertilizer applicator consists of two core control components: rotational adjustment of the input shaft (speed regulation mechanism) and axial displacement (directional switching mechanism). By adjusting the circumferential rotation speed of the input shaft, the fertilizer amount discharged per revolution of the grooved wheel can be modulated. Meanwhile, axial displacement of the input shaft changes the effective working length of the grooved wheel, thus enabling dual regulation of fertilizer discharge.

2.2.4.1 Design of the Switching Mechanism

Unlike conventional manual or semi-automatic adjustment structures, the proposed switching mechanism transforms the rotary motion of a stepper motor into linear displacement via a lead screw–nut pair. This linear movement is further transmitted to the grooved wheel axially via a lever mechanism. The schematic diagram of the switching mechanism is shown in Fig. 2.6.

A simplified force analysis was conducted on the lever. For equilibrium, the torque generated by the driving force and resistance force must satisfy the balance condition:

$$F_a \times OC = F_b \times OD \tag{2.1}$$

Where is the driving force exerted on the nut, and is the resistance force on the main shaft slider. OC and OD are the respective moment arms. In the similar triangles triangle△AOC and triangle△BOD, the geometric relationships can be derived to link nut displacement with the axial displacement of the grooved wheel. Therefore, by properly positioning the pivot point, the axial travel of the groove wheel can be accurately inferred from the nut's displacement, facilitating precise control of the fertilizer dispensing width.

$$\frac{F_a}{F_b} = \frac{OD}{OC} \tag{2.2}$$

$$\frac{DB}{CA} = \frac{OD}{OC} \tag{2.3}$$

$$CA = \frac{OC \cdot DB}{OD} \tag{2.4}$$

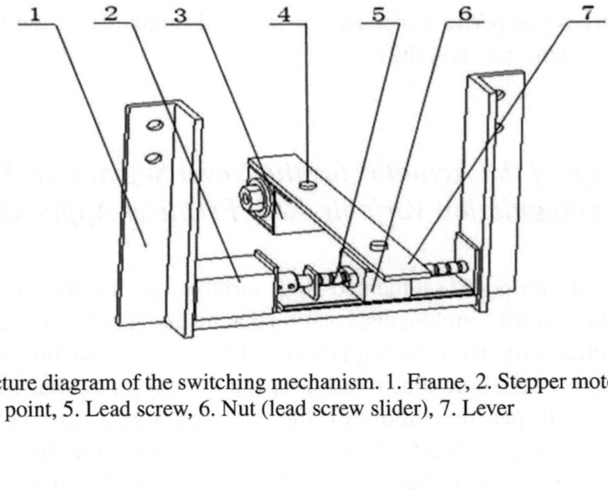

Fig. 2.6 Structure diagram of the switching mechanism. 1. Frame, 2. Stepper motor, 3. Main shaft slider, 4. Pivot point, 5. Lead screw, 6. Nut (lead screw slider), 7. Lever

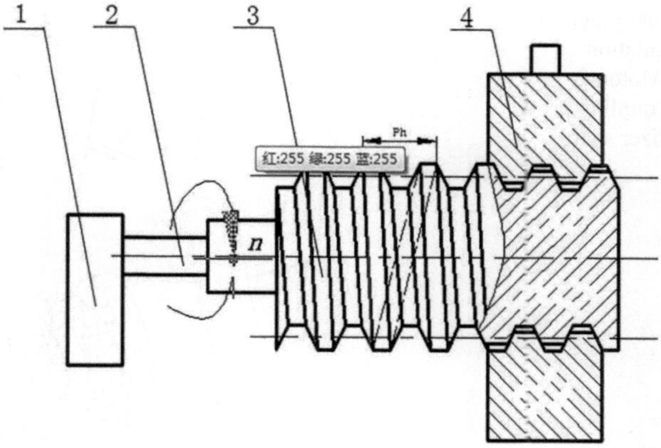

Fig. 2.7 Structure diagram of screw drive. 1. Stepper motor, 2. Motor shaft, 3. Lead screw, 4. Nut (lead screw slider)

2.2.4.2 Analysis of the Lead Screw Transmission

In the helical drive structure, the lead screw and nut form a pair that converts rotational motion into linear displacement. For each full rotation, the nut moves linearly by the lead ppp. Thus, the displacement sss is:

$$L = N \cdot P_h \tag{2.5}$$

Where

L—linear travel distance of the moving element, mm
N—number of motor revolutions, r
P_h—lead of the thread, mm

The designed actuator utilizes a single-start right-hand standard thread as shown in Fig. 2.7. If the lead ppp and the motor speed P_h are known, then within a unit time n, the axial displacement of the nut can be calculated by:

By integrating the screw drive and lever mechanism, the axial displacement ΔL\ Delta LΔL of the main shaft slider can be expressed as a function of motor input:

$$L_{\neq} = \frac{CO \cdot BD}{DO} = \frac{CO \cdot (n \cdot P_h)}{60 \times DO} \tag{2.6}$$

This relation indicates that the groove wheel's axial movement can be precisely determined based on the motor's rotational input, thus enabling accurate real-time control of fertilizer flow and achieving effective variable-rate fertilization.

2.2.4.3 Design of the Speed Regulation Mechanism

The speed regulation mechanism adopts a direct-drive configuration, where a motor directly drives the fertilizer dispenser shaft. The structural schematic is shown in Fig. 2.8.

Fig. 2.8 Structure diagram of the speed regulation mechanism. 1. Motor, 2. Side panel, 3. Coupling device, 4. Fertilizer shaft

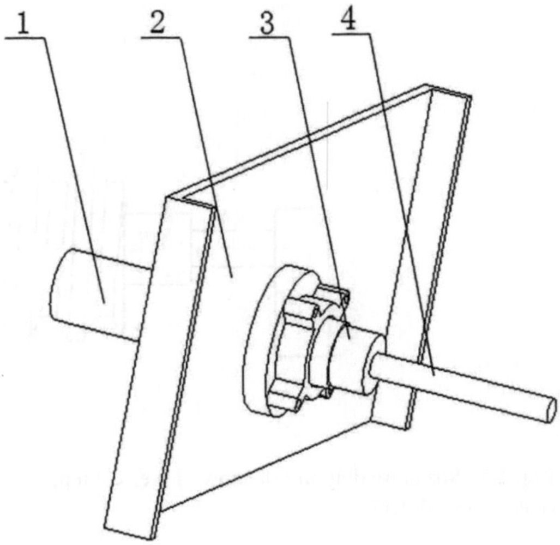

Assuming negligible nonlinearity due to shaft acceleration, the fertilizer application rate for a single dispenser is linearly related to shaft speed:

$$q = k \cdot n + b \quad (2.7)$$

Where

q—fertilizer output rate, g/min
n—shaft speed, r/min
k—slope of fitted line (calibration coefficient)
b—intercept of fitted line

It should be noted that the k and b values are specific to each dispenser configuration. Therefore, calibration experiments must be conducted for every fertilizer dispensing unit to ensure precision in variable-rate control.

2.3 Development of Discrete Element Model and Simulation Design for Variable-Rate Fertilizer Applicator

2.3.1 Fundamental Theory of Discrete Element Simulation

The Discrete Element Method (DEM) is a numerical approach widely used to simulate and analyze granular systems. It treats contacting bodies as discrete particles and considers factors such as particle shape, physical properties, and size distribution. Based on Newton's second law, it calculates forces, positions, and velocities of individual particles by solving parameterized equations of motion within a granular assembly [1, 27]. The DEM simulation involves two primary components: contact force models and Newtonian mechanics.

According to the force-displacement relationship, forces can be derived from particle displacements, which are obtained by solving Newton's second law:

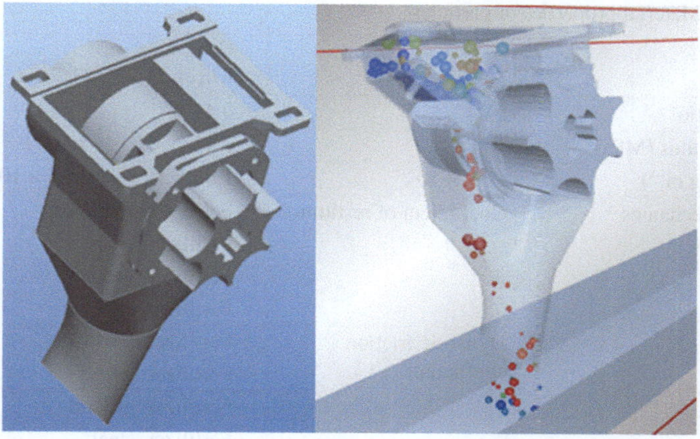

Fig. 2.9 Assembly and simulation drawing of fertilizer apparatus

$$\begin{cases} \sum F = m_i \ddot{u}_i \\ \sum M = I_i \ddot{\theta}_i \end{cases} \tag{2.8}$$

2.3.2 DEM Model Construction and Material Parameter Selection

Following standard EDEM modeling procedures, a 3D model of the outer-grooved wheel fertilizer dispenser was developed using Pro/E. The model was assembled with virtual constraints and exported in *.igs format to EDEM for simulation. The structural model and corresponding simulation configuration are shown in Fig. 2.9.

Commercial granular compound fertilizer commonly used in southern China was selected as the simulation material. To simplify the modeling, particles were treated as ideal spheres. Material property parameters were derived from literature and experimental sampling. The contact interactions were modeled using the default Hertz-Mindlin (no-slip) model in EDEM. Simulation settings are provided in Table 2.1, and the total number of particles in the virtual environment was 100,000. The combined rotational and translational motion of the applicator was configured to replicate realistic field operations.

2.3.3 Simulation Design for Fertilizer Dispensing Process

The three types of improved grooved-wheel dispensers (shown in Fig. 2.8.) were imported into EDEM for comparative simulations. Each test configuration was repeated five times, and the average fertilizer mass on the virtual ground was used to evaluate consistency via the coefficient of variation (CV).

Table 2.1 Material properties of particles

Parameter	Fertilizer	Groove wheel	Ground	Steel
Poisson ratio	0.24	0.47	0.51	0.3
Shear modulus (MPa)	10.76	0.95	1.1×10^2	7×10^4
Density (kg m^{-3})	1.38×10^3	1.06×10^3	1.25×10^3	7.8×10^3
Contact mechanics parameters	Coefficient of restitution	Fertilizer–fertilizer		0.27
		Fertilizer–wheel		0.35
		Fertilizer–ground		0.06
		Fertilizer–steel		0.29
	Dynamic friction coefficient	Fertilizer–fertilizer		0.26
		Fertilizer–wheel		0.32
		Fertilizer–ground		1.20
		Fertilizer–steel		0.59
	Static friction coefficient	Fertilizer–fertilizer		0.34
		Fertilizer–wheel		0.47
		Fertilizer–ground		1.13
		Fertilizer–steel		0.48
Particle diameter (mm)	3.96			

A dual-variable control strategy was employed by altering the groove opening and wheel speed. In the *speed-priority* control mode, the opening was adjusted at fixed rotational speeds to identify the optimal setting, which was then held constant while varying speed for control. Conversely, in the *opening-priority* mode, the groove opening remained fixed while wheel speeds were varied. Simulation outputs were used to calculate the fertilizer discharge stability for each control strategy.

The coefficient of variation (CV), a key indicator of dispensing uniformity, was calculated using:

$$Cv = \frac{\sigma}{\overline{m}} \times 100\% \tag{2.9}$$

2.3.4 Simulation Results and Analysis

2.3.4.1 Structural Simulation Analysis of Fertilizer Applicators

Based on the previously described experimental scheme, the fertilizer dispensing performance of grooved-wheel dispensers with different structural configurations was evaluated under the same operating parameters (opening and rotating speed). The statistical results of the discharge standard deviation and coefficient of variation (CV) are shown in Fig. 2.10.

As illustrated in Fig. 2.10, compared with the conventional grooved-wheel dispenser (type a), the improved structures (types b and c) exhibited significantly enhanced performance. For type a, the standard deviation of discharge ranged from

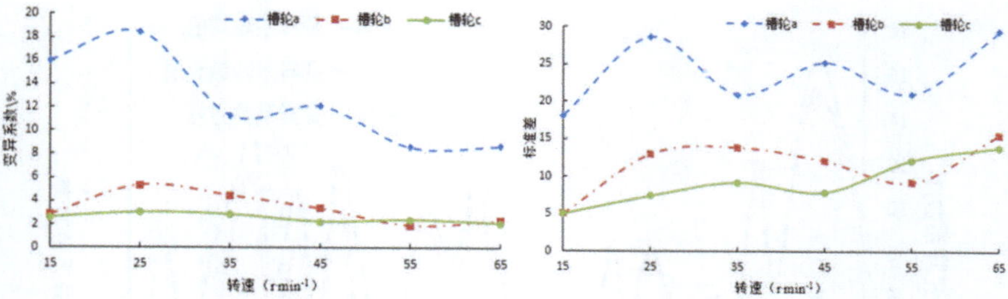

Fig. 2.10 Simulation results of fertilizer discharge among different grooved-wheel structures

18.09 to 29.39, with an average of 23.88 g. The corresponding CV ranged from 8.57% to 18.39%, averaging 12.49%. In contrast, type b had a much lower average standard deviation of 11.47 g and an average CV of 3.39%, while type c showed the best performance with an average standard deviation of 9.29 g and CV of 2.53%.

Compared with type a, the average standard deviation and CV for type b decreased by 51.97% and 72.86%, respectively; for type c, the decreases were 61.1% and 79.74%. Thus, the optimized dispenser structure (type c) achieved the most stable fertilizer output, with a maximum CV of only 3.01%, demonstrating significantly improved discharge uniformity for variable-rate fertilization.

2.3.5 *Simulation Analysis of Control Strategies*

2.3.5.1 Simulation Analysis of Speed-Priority Control

The simulation analysis of fertilizer discharge variation under different groove openings, using a speed-priority control strategy, is shown in Fig. 2.11. The maximum CV reached 15.25%, with a maximum mean CV of 7.08%, a minimum of 0.97%, and a maximum variance of 2.32%.

When the groove opening was set to 45.16 mm, the CV maximum, mean, and variance reached their respective minimum values of 1.94%, 0.97%, and 0.31. Considering mechanical constraints in the actual adjustment mechanism, to avoid issues such as jamming or disconnection due to excessive openings, a more practical opening of 19.34 mm was selected. At this setting, the CV metrics slightly increased to 1.98% (max), 1.09% (mean), and 0.38 (variance), but remained acceptably low. Therefore, in speed-priority mode, maintaining a constant opening of 19.34 mm provides near-optimal discharge stability.

2.3.5.2 Simulation Analysis of Opening-Priority Control

Figure 2.12 presents the CV statistics when groove rotational speed is varied under an opening-priority control strategy. The maximum CV reached 10.49%, with a maximum mean of 3.81%, a minimum mean of 1.84%, and a maximum variance of 4.17.

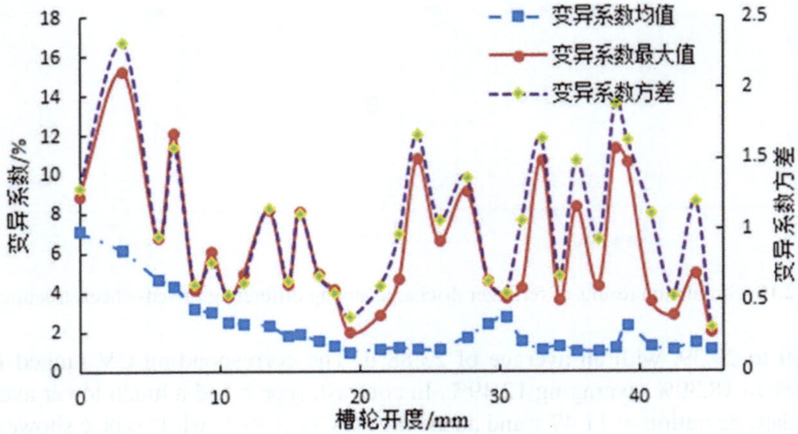

Fig. 2.11 Variation coefficients under different groove openings

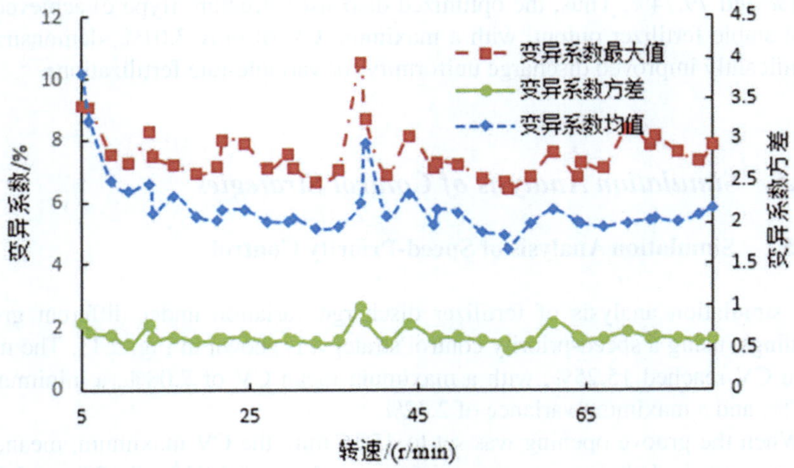

Fig. 2.12 Variation coefficients under different rotating speeds

Compared to the speed-priority strategy, opening-priority control resulted in slightly higher CVs, but demonstrated more consistent discharge performance, as indicated by the lower variance fluctuation. The optimal performance was observed at a groove speed of 55.75 r/min, where the CV maximum, mean, and variance dropped to 6.25%, 1.85%, and 1.42, respectively. Hence, for stable operation in opening-priority mode, a fixed groove speed of 55.75 r/min is recommended.

Through integration with a crop canopy NDVI-based decision system, the control system of the rice-wheat variable-rate fertilizer applicator calculates the target fertilizer rate in real time. By combining this with the current travel speed, it dynamically determines the optimal groove opening and rotation speed, achieving high-precision variable-rate fertilization.

2.4 Fertilization Strategy Optimization for Rice-Wheat Variable Rate Applicator Based on Multilayer Perceptron Model

2.4.1 Fertilizer Discharge Prediction Model Based on LFA-PSO-MLP Algorithm

2.4.1.1 Principle of Particle Swarm Optimization (PSO)

The Particle Swarm Optimization (PSO) algorithm is a swarm intelligence optimization technique inspired by the social behavior of birds and fish. In PSO, each individual, referred to as a "particle", represents a candidate solution in the search space. These particles adjust their positions by learning from both their individual best experience and the global best experience of the swarm.

For the *i*th particle, the velocity and position update equations are as follows:

$$v_i^{(t+1)} = \omega \cdot v_i^{(t)} + c_1 \cdot r_1 \cdot \left(P_i^{(t)} - x_i^{(t)}\right) + c_2 \cdot r_2 \cdot \left(P_g^{(t)} - x_i^{(t)}\right) \tag{2.10}$$

$$x_i^{(t+1)} = x_i^{(t)} + v_i^{(t+1)} \tag{2.11}$$

where ω is the inertia weight; c_1 and c_2 are acceleration coefficients; r_1 and r_2 are random numbers within [0, 1]; $P_i^{(t)}$ is the personal best position; $P_g^{(t)}$ is the global best position; $x_i^{(t)}$ $v_i^{(t+1)}$ represent the current position and velocity of the particle.

Although PSO exhibits strong global optimization capabilities, it is sensitive to parameter settings and prone to premature convergence [28, 29]. To address these issues, we introduce the Levy Flight Algorithm to enhance the exploration ability and randomness of the particles [30].

2.4.1.2 Principle of Levy Flight Algorithm (LFA)

The Levy Flight Algorithm (LFA) is a stochastic optimization method characterized by long-tailed step-length distributions. Inspired by foraging patterns of animals such as birds and insects, LFA introduces large, infrequent jumps in particle movement, increasing the likelihood of escaping local optima.

The step length in Levy flights follows a probability density function defined by:

$$p(x) = \frac{1}{\sqrt{2\pi\sigma^2}} e^{-\frac{1}{2}\left(\frac{1}{x}\right)^{\frac{1}{\beta}}} \tag{2.12}$$

where $\beta \in (0,2]$ is the shape parameter and sss is the step size. The position update rule is given by:

$$x_{new} = x_{old} + \alpha \cdot Levy \tag{2.13}$$

where alpha (α) is a scaling. By integrating LFA into PSO, we enhance the stochasticity and diversity of particle trajectories, significantly improving the global search capability and reducing the risk of stagnation in local optima.

2.4.1.3 Principle of Multilayer Perceptron (MLP)

The Multilayer Perceptron (MLP) is a type of feedforward neural network composed of an input layer, one or more hidden layers, and an output layer. Each neuron in the network computes a weighted sum of its inputs, adds a bias, and applies a nonlinear activation function:

$$z_j = \sum_{i=1}^{n} w_{ij} \cdot x_i' + b_j \tag{2.14}$$

For the output neuron kkk, the weighted sum and final output are calculated as:

$$a_j = f\left(z_j\right) \tag{2.15}$$

$$z_k = \sum_{j=1}^{m} w_{jk} \cdot a_j + b_k \tag{2.16}$$

MLP is capable of approximating complex nonlinear functions and is particularly suited for regression tasks such as fertilizer rate prediction [31].

Finally, the weighted sum of inputs is passed through the activation function in the output layer to produce the output of the output neurons. When extending from a single hidden layer to multiple hidden layers, the process involves sequentially applying weight multiplication, bias addition, and activation functions from the input layer through each hidden layer. The output of one layer serves as the input to the next, ultimately yielding the final output at the output layer. The network architecture of the MLP model is illustrated in Fig. 2.13.

By incorporating the LFA-optimized PSO algorithm into the MLP framework, the model's generalization capability is significantly enhanced, enabling it to better adapt to diverse and complex datasets. This integration not only improves the global optimization ability of the model but also increases computational efficiency, shortens training time, and enhances the overall fitting and predictive performance of the model.

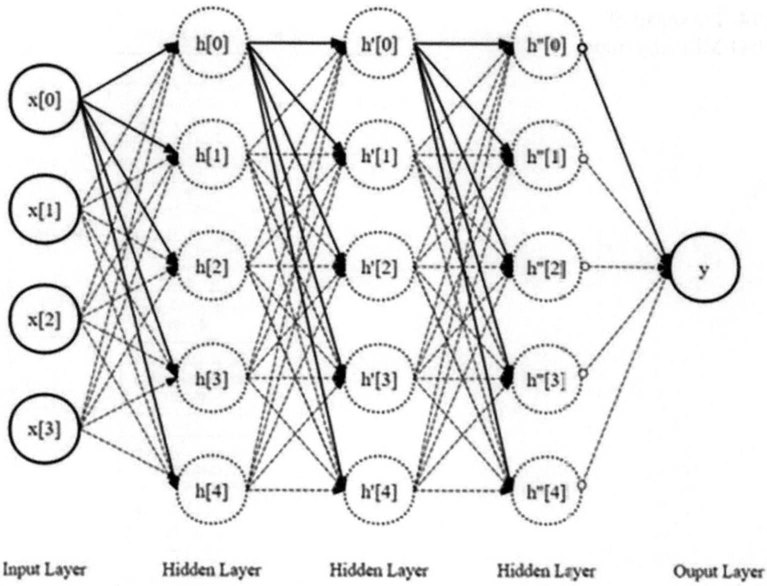

Fig. 2.13 The Network Structure of the MLP Model

2.4.1.4 Construction Method of the LFA-PSO-MLP Fertilizer Application Prediction Model

The algorithm flowchart is illustrated in Fig. 2.14. The entire process consists of five key stages: algorithm parameter initialization, dataset preparation, algorithmic optimization, model optimization, and model testing. Initially, the parameters for LFA, PSO, and MLP are independently configured. The dataset is then divided into training and testing subsets according to a defined ratio, followed by normalization to eliminate dimensional inconsistencies and reduce bias among feature scales. Subsequently, the characteristic parameters of individual PSO particles are updated, and the LFA algorithm is employed to refine particle positions and velocities, thereby obtaining the optimal prediction parameters. These parameters are used to initialize and train the MLP model. To enhance generalization and prevent overfitting, L2 regularization is incorporated. The result is an optimized predictive model capable of accurately estimating fertilizer application rates.

2.4.2 Fertilizer Application Calibration Experiment

To establish a relational model among fertilizer shaft opening, rotational speed, and fertilizer discharge rate for achieving precise fertilizer control, a calibration experiment was conducted using granular compound fertilizer commonly applied in the

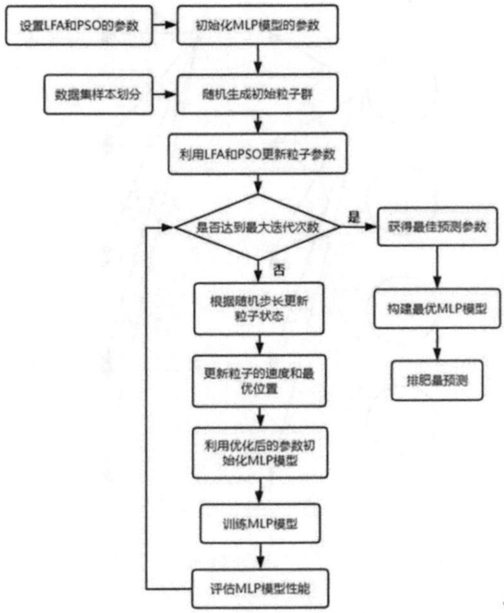

Fig. 2.14 Flowchart of LFA-PSO-MLP algorithm

rice–wheat rotation regions of the middle and lower reaches of the Yangtze River, particularly Jiangsu Province. The experiment was carried out at the Intelligent Agricultural Machinery Laboratory, College of Engineering, Nanjing Agricultural University. The fertilizer applicator used in the experiment was a dual-variable fertilizer device developed by the research group (as shown in Fig. 2.15).

To meet the operational requirements for adjusting the fertilizer outlet opening and shaft rotational speed, parameter gradients were defined within their respective adjustment ranges. The fertilizer outlet opening was set from 20 to 45 mm with an increment of 5 mm, while the shaft rotational speed was set from 20 to 60 r/min with an increment of 5 r/min, resulting in a total of 54 combinations of opening-speed pairs. The coding levels of the experimental variables are shown in Table 2.2, where the upper and lower bounds of the opening and speed adjustment ranges were assigned level values of 1 and −1, respectively. Other combinations within the range were defined using a stepwise gradient design. Each combination was tested for a duration of 1 min, with three repetitions, and the average fertilizer discharge from the three trials was recorded as the final result.

The fertilizer discharge data obtained from the calibration experiments were analyzed using Design-Expert 10.0.1 software to establish a response model describing the relationship between fertilizer outlet opening (L), shaft rotational speed (N), and fertilizer discharge rate (Q). The single-factor response curves and the interaction response surface are shown in Fig. 2.16, while the normal distribution of externally studentized residuals is presented in Fig. 2.17.

Fig. 2.15 Indoor calibration test site

Table 2.2 Experimental variable encoding level table

Test variables		Coded value		Interval
		−1	1	
A	Opening *L* (mm)	20	45	5
B	Speed *N* ($r \cdot min^{-1}$)	20	60	5
C	Fertilizer amount *Q* ($g \cdot min^{-1}$)	/		

2.4.2.1 Analysis of Response Curves and Surfaces

As shown in Fig. 2.16, under single-factor influence, the fertilizer discharge rate increases linearly with both the outlet opening and shaft rotational speed. The discharge rate follows a near-linear trend with the minimum and maximum values occurring at the −1 and +1 coded levels of opening and speed, respectively. It is evident that the shaft rotational speed has a more significant effect on the fertilizer discharge rate than the outlet opening.

Furthermore, the interaction between the opening and rotational speed exerts a greater influence on the discharge rate than either factor alone. Nevertheless, the overall trend remains a steady increase. The minimum and maximum discharge rates were observed at interaction levels of −1 and +1, corresponding to 656.48 and 4606.55 g/min, respectively.

2.4.2.2 Analysis of Externally Studentized Residuals

Externally Studentized Residuals (ESR) are commonly used to identify outliers in regression analysis. The ESR is calculated by dividing each residual by its standard error, which helps to determine the degree to which a data point deviates from the model's prediction. In general, an observation with an absolute ESR greater than 3

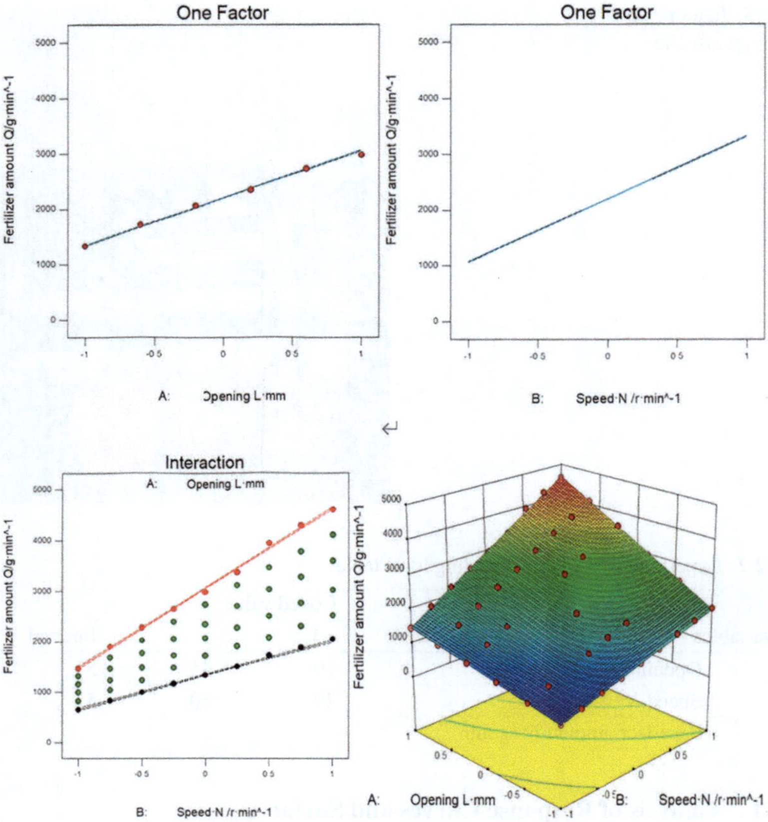

Fig. 2.16 Effects of single and interaction factors on fertilizer distribution

is considered an outlier, while values with an absolute ESR less than 2 are regarded as acceptable and reliable for analysis.

As shown in Fig. 2.17, among the 54 opening-speed combinations tested, only 3 exhibited ESR values close to ±3, indicating potential outliers. The remaining 51 combinations had ESR values within the range of [0, 2], suggesting that the residual distribution was largely normal and the data quality was robust. These results confirm the reliability of the calibration experiment, indicating that the collected data can be effectively used as a reference for modeling fertilizer discharge rates.

2.4.3 *Construction and Parameter Optimization of the LFA-PSO-MLP Model*

The development and execution environment for the LFA-PSO-MLP model algorithm was configured on a system with an Intel(R) Core(TM) i5-9300H CPU @ 2.40 GHz, 16 GB RAM, running a 64-bit Windows 10 operating system. The

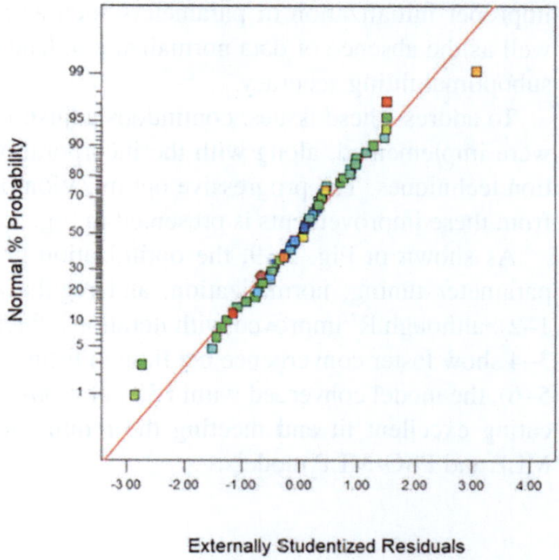

Fig. 2.17 Normal distribution of externally studentized residuals

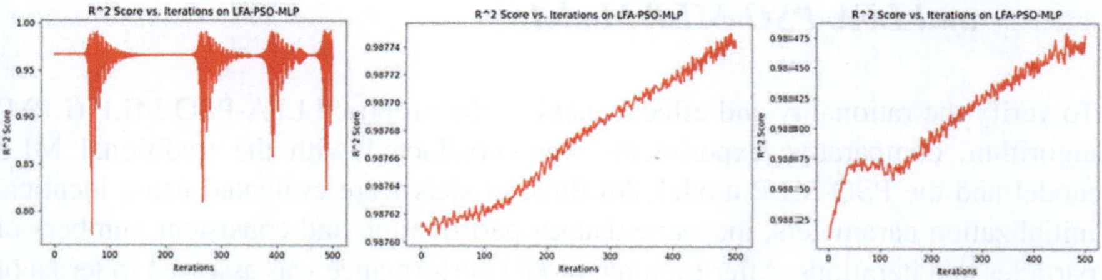

Fig. 2.18 Preliminary Model of LFA-PSO-MLP

programming and testing were conducted using Visual Studio Code version 1.88.1.

Based on the methodology described in Sect. 2.4, the LFA-PSO-MLP model (hereafter abbreviated as LPM unless otherwise specified) was initially constructed. The model parameters were configured as follows: for the Particle Swarm Optimization (PSO) component, the particle population was set to 10, with an initial position vector of [0.05, 0.05] and a maximum of 100 iterations. For the Levy Flight Algorithm (LFA), the Levy distribution parameter was set to 1.5, with a flight step size range of [0, 0.1], an inertia weight of 1.0, and learning factors both set to 1.0. For the Multilayer Perceptron (MLP) component, the minimum learning rate was configured to 0.001, and the maximum number of training iterations was set to 500.

The coefficient of determination (R^2) curve obtained from the model training process is illustrated in Fig. 2.18, indicating the model's fitting performance.

As illustrated in Fig. 2.18, the R^2 fitting curves of the three training iterations exhibit significant oscillations, with some trajectories even diverging outside the convergence domain. The final R^2 values obtained after training remain low and fail to meet the expected performance criteria. This instability can be attributed to the improper initialization of parameters such as inertia weight and learning rate, as well as the absence of data normalization, leading to poor model convergence and suboptimal fitting accuracy.

To address these issues, continuous adjustments to the initialization parameters were implemented, along with the incorporation of normalization and regularization techniques. The progressive optimization process of the LPM model resulting from these improvements is presented in Fig. 2.19.

As shown in Fig. 2.19, the optimization process of the LPM model involved parameter tuning, normalization, and regularization. In early stages (subfigures 1–2), although R^2 improved with iterations, fluctuations remained large. Subfigures 3–4 show faster convergence but limited fitting accuracy. In later stages (subfigures 5–6), the model converged within 50 iterations, achieving R^2 values close to 1, indicating excellent fit and meeting the requirements for comparative analysis with MLP and PSO-MLP models.

2.4.4 Comparison of MLP, PSO-MLP, and LFA-PSO-MLP Models

To verify the rationality and effectiveness of the proposed LFA-PSO-MLP (LPM) algorithm, comparative experiments were conducted with the traditional MLP model and the PSO-MLP model. All three models were evaluated using identical initialization parameters, the same dataset partitioning, and consistent numbers of particles and iterations. After training, model performance was assessed in terms of convergence behavior and fitting accuracy.

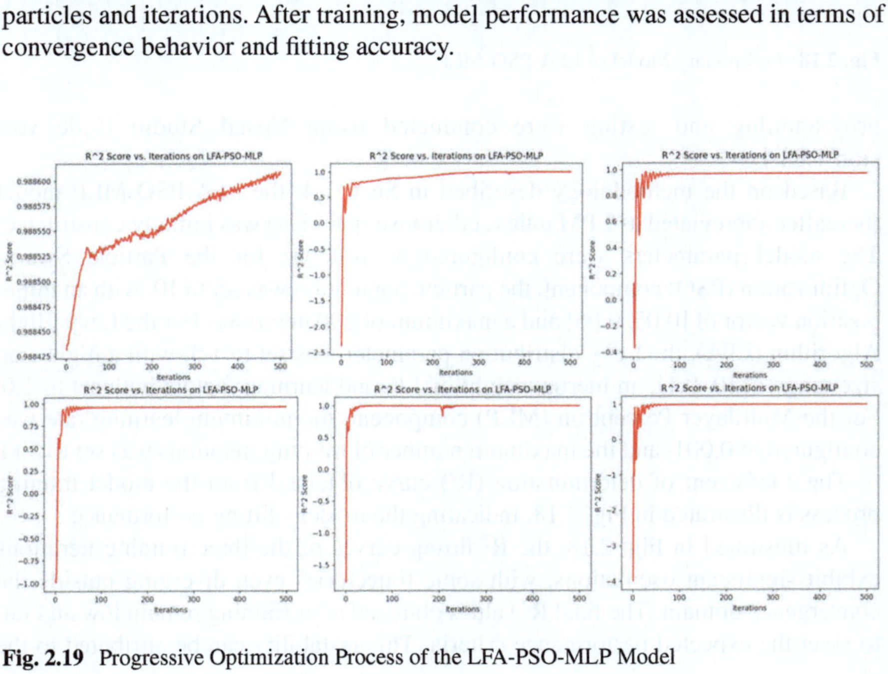

Fig. 2.19 Progressive Optimization Process of the LFA-PSO-MLP Model

2.4.4.1 Experimental Environment and Parameter Settings

To ensure experimental validity and fairness, all three models were trained and tested on the same dataset, which consisted of 205 groups of test data. The fertilizer shaft opening and rotation speed were used as input features, and the fertilizer amount served as the target output. The dataset was split into training and testing sets at a ratio of 0.85:0.15. For initialization, the PSO particle count was set to 10 (not applicable to models without PSO), the initial position to [0.05, 0.05], and the maximum number of iterations to 100. The LFA parameters were set as follows: Lévy flight index = 2, step size range = [0, 0.01], inertia weight = 0.5, and cognitive and social learning factors = 1.5. The MLP model was configured with a minimum learning rate of 0.0001, a regularization parameter of 0.05, and a maximum of 500 iterations.

To mitigate the influence of random anomalies on the final conclusions, each model was trained ten times for 500 iterations per run. The most frequent result among these trials was taken as the representative output. If outlier results were frequently observed, additional training rounds were conducted to ensure result robustness and reproducibility.

2.4.4.2 Results and Analysis

The convergence curves for the three models during training are shown in Fig. 2.20.

In Fig. 2.20, the MLP model demonstrated slow convergence within the first 100 iterations, followed by rapid fitting between iterations 100 and 300, and then gradually leveled off, with R^2 steadily approaching 1. The PSO-MLP model maintained an R^2 range of [0.975, 1] throughout training but exhibited significant oscillations, making it difficult to achieve stable convergence. This instability is attributed to the large parameter space of the MLP model and the presence of multiple local optima, causing PSO to jump between them. In contrast, the LPM model showed rapid and stable convergence within the first 50 iterations. After a brief period of minor R^2 fluctuations, it transitioned to a stable state, consistently approaching $R^2 \approx 1$.

Each model was trained ten times. The stable iteration number, final R^2 value at 500 iterations, and mean absolute percentage error (MAPE) are summarized in

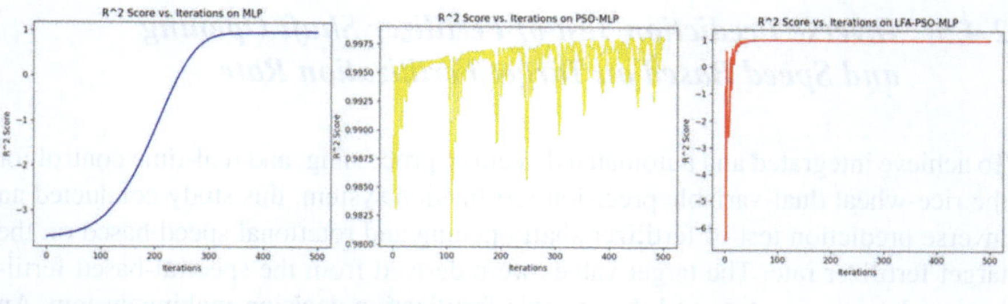

Fig. 2.20 Convergence curves of R^2 for three models

Table 2.3 R^2 and MAPE obtained from model training

Number of training iterations	MLP		PSO-MLP		LFA-PSO-MLP	
	R^2	*MAPE*	R^2	*MAPE*	R^2	*MAPE*
1	0.943	13.01	0.992	2.87	0.998	1.93
2			0.993	2.79	0.999	1.80
3			0.989	2.99	0.998	1.83
4			0.995	2.80	0.998	1.89
5			0.987	3.08	0.999	1.77
6			0.989	3.02	0.999	1.80
7			0.991	2.94	0.998	1.84
8			0.988	2.99	0.998	1.84
9			0.99	2.96	0.999	1.79
10			0.992	2.92	0.998	1.84
Mean value			0.988	2.95	0.999	1.83
Stable fitting iteration	300		/		50	

Table 2.3. Notably, the MLP model achieved optimal results within 500 iterations across all runs; hence, its R^2 and MAPE remained constant. The PSO-MLP and LPM models did not reach full convergence within 500 iterations. However, their test accuracy at 500 iterations was sufficiently high, and the time cost of further training exceeded the marginal gains in accuracy. Therefore, the maximum iteration number was fixed at 500.

In summary, the MLP model exhibited gradual convergence with limited prediction accuracy (R^2 = 0.943, MAPE = 13.01%). The PSO-MLP model showed improved accuracy but with significant fluctuation. The LPM model demonstrated the best overall performance, with faster convergence, higher stability, and superior fitting capability. Although the difference between PSO-MLP and LPM in R^2 was only 0.011 and in MAPE 1.12%, the convergence smoothness and model robustness of LPM were notably better (Table 2.4 and Fig. 2.21).

Therefore, the LFA-PSO-MLP model is validated as a more reliable and effective prediction tool for estimating fertilizer output based on actuator opening and shaft speed, and is adopted in this study for constructing the fertilizer rate prediction model.

2.4.5 *Inverse Prediction Test of Fertilizer Shaft Opening and Speed Based on Target Fertilization Rate*

To achieve integrated and automated detection, processing, and real-time control for the rice-wheat dual-variable precision fertilization system, this study conducted an inverse prediction test of fertilizer shaft opening and rotational speed based on the target fertilizer rate. The target values were derived from the spectral-based fertilization detection system and the variable fertilization decision-making system. An

Table 2.4 Comparison of validation and predicted values

Serial number	Feature variable Opening L (mm)	Speed N (r·min^{-1})	Verification value Q (g·min^{-1})	Predicted value P (g·min^{-1})	Relative error (%)
1	23	23	1020.438	1034.406	1.37
2	23	33	1390.638	1419.156	2.05
3	23	43	1751.322	1821.186	3.99
4	23	53	2151.618	2213.298	2.87
5	23	63	2504.772	2595.048	3.6
6	35	23	1435.152	1428.222	0.48
7	35	33	2044.548	2003.796	1.99
8	35	43	2627.13	2559.702	2.57
9	35	53	3212.91	3156.564	1.75
10	35	63	3829.41	3721.314	2.82
11	43	23	1866.858	1824.606	2.26
12	43	33	2642.034	2556.012	3.26
13	43	43	3415.578	3322.296	2.73
14	43	53	4182.63	4065.072	2.81
15	43	63	4911.174	4791.09	2.45
Mean value (%)					2.47

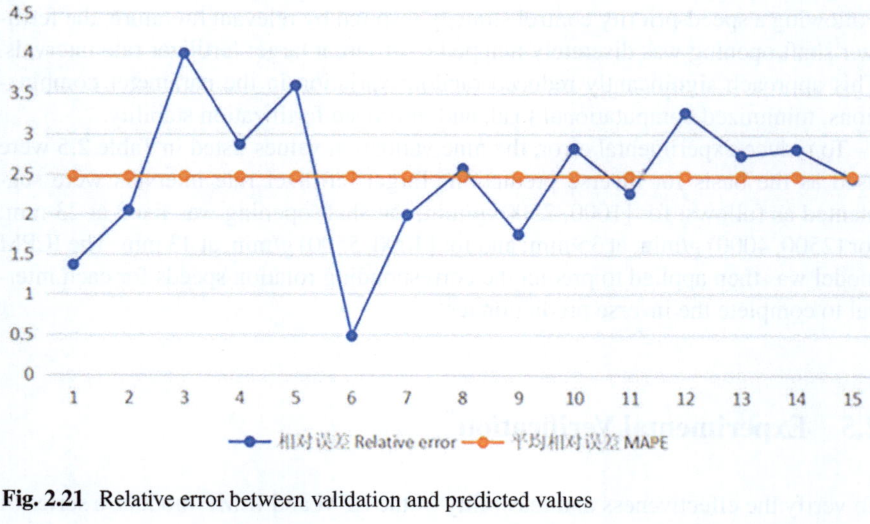

Fig. 2.21 Relative error between validation and predicted values

Inverse LFA-PSO-MLP (ILPM) model was developed using Visual Studio Code 1.88.1, based on the previously constructed optimal LFA-PSO-MLP (LPM) forward prediction model.

The results are shown in Table 2.5. The maximum, minimum, and average relative errors between the reference and predicted rotation speeds were 2.52%, 1.06%,

Table 2.5 Data related to inverse prediction of the ILMP model

Serial number	Verification value Q (g·min^{-1})	Opening *L* (mm)	Reference value *N* (r·min^{-1})	Predicted value N_p (r·min^{-1})	Relative error (%)
1	1020.438	23	23	23.53	2.30
2	1390.638	23	33	32.48	1.58
3	1751.322	23	43	43.96	2.23
4	2151.618	23	53	54.01	1.91
5	2627.13	35	43	42.39	1.42
6	3212.91	35	53	53.64	1.21
7	3829.41	35	63	64.59	2.52
8	4182.63	43	53	54.12	2.11
9	4911.174	43	63	62.33	1.06
Mean value /%					1.82

and 1.82%, respectively. These results demonstrate that the ILPM model can effectively determine appropriate combinations of shaft opening and rotational speed for a given target fertilizer rate, thereby providing a solid theoretical foundation for the seamless integration of the spectral detection module and the variable-rate fertilization decision system in the rice-wheat dual-variable precision fertilizer applicator.

Since multiple combinations of shaft opening and rotational speed can correspond to the same target fertilizer rate, a gradient-based control strategy was designed to constrain the solution space and improve computational efficiency. Following a speed-priority control strategy inspired by relevant literature, the fertilizer shaft opening was discretely mapped to different target fertilizer rate intervals. This approach significantly reduced random variation in the parameter combinations, minimized computational load, and improved fertilization stability.

To reduce experimental error, the nine validation values listed in Table 2.5 were used as the basis for inverse prediction. Target fertilizer rate intervals were segmented as follows: for [1000, 2500) g/min, the shaft opening was fixed at 23 mm; for [2500, 4000) g/min, at 35 mm; and for [4000, 5500) g/min, at 43 mm. The ILPM model was then applied to predict the corresponding rotation speeds for each interval to complete the inverse prediction test.

2.5 Experimental Verification

To verify the effectiveness and feasibility of the LPM and ILMP models, a series of tests were conducted using the self-developed dual-variable precision fertilizer applicator for rice and wheat, designed by the research team. These tests were carried out at the Wulie Modern Agricultural Demonstration Park in Dongtai, Jiangsu Province. The test setup is shown in Fig. 2.22.

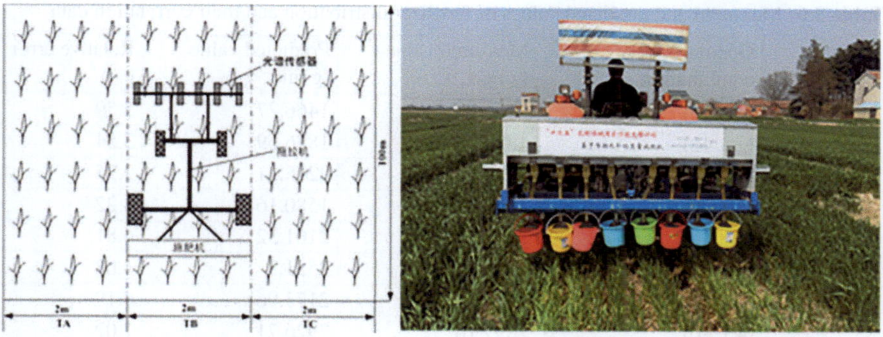

Fig. 2.22 Fertilizer distribution test verification site

2.5.1 Fertilizer Application Rate Prediction Test Based on LFA-PSO-MLP

The traveling speed of the fertilizer applicator was set to 1 m/s, with the opening degree of the fertilizer dispenser ranging from 20 to 45 mm and the rotational speed ranging from 20 to 60 r/min. The fertilizer used in the experiment was a compound fertilizer developed by Nanjing Zhengmei Shinnong Chemical Co., Ltd. A total of 10 fertilization tests were carried out under the dual-variable conditions of opening degree and rotational speed, each lasting for 1 min. During each test, the actual values of the opening degree, rotational speed, and fertilizer application rate were recorded.

These actual measurements were compared with the fertilizer application rates predicted by the LPM model. The relative error (RE) was calculated using the formula: The measured and predicted fertilizer application rates along with the corresponding relative errors are shown in Table 2.6. The fertilizer rate prediction performance of the LPM model is illustrated in Fig. 2.23.

In Table 2.6, among 10 different combinations of opening degree and rotational speed, the measured and predicted fertilizer application rates ranged from 1388.74 to 4137.22 g/min and 1466.27 to 4094.76 g/min, respectively, both meeting the standard requirements for normal fertilizer distribution. The maximum relative error between the measured and predicted values was 5.59%, the minimum was 1.02%, and the overall average relative error was 3.49%, indicating relatively low deviation and thus high prediction accuracy.

Figure 2.23 further shows that the predicted values obtained from the test closely fluctuate around the same level as the actual values, with only minor differences. This demonstrates that the model fits the real fertilizer application data well. The prediction results meet the expected outcomes of the experiment and can serve as a preliminary validation of the ILMP prediction model.

Table 2.6 Measured and predicted values of fertilizer distribution and their correlation data

Serial number	Opening, speed combination	Measured value (g·min^{-1})	Predicted value (g·min^{-1})	Relative error (%)
1	(25, 30)	1388.74	1466.27	5.59
2	(25, 40)	1822.21	1886.69	3.54
3	(25, 50)	2330.98	2276.31	2.35
4	(30, 30)	1649.75	1580.16	4.22
5	(30, 40)	2207.43	2101.32	4.81
6	(30, 50)	2431.96	2531.3	4.08
7	(40, 30)	2269.11	2173.96	4.19
8	(40, 40)	2897.16	2926.71	1.02
9	(45, 40)	3347.4	3210.62	4.09
10	(45, 50)	4137.22	4094.76	1.03
MAPE (%)				3.49

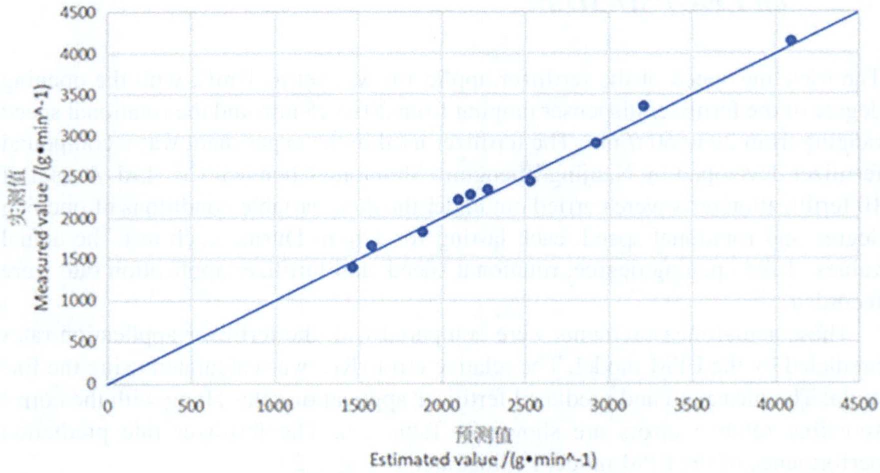

Fig. 2.23 Fertilizer distribution prediction performance based on the LFA-PSO-MLP model

2.5.2 *Inverse LFA-PSO-MLP Opening Degree and Rotational Speed Prediction Test*

After verifying the feasibility and effectiveness of the LMP model, further tests were carried out in the park using a fertilizer applicator equipped with the ILMP model. In this test, the applicator traveled in a straight line at a speed of 1 m/s for a total of 6 runs, each covering an operational area of 180 m^2. For each run, the target fertilizer application rate was recorded, along with the actual fertilizer amount achieved through real-time adjustment of opening degree and rotational speed by the ILMP model.

The relative error between the target and actual application rates was used as the evaluation metric to assess the fertilization performance of the applicator equipped with the ILMP model. The test results are shown in Table 2.7.

In Table 2.7, the relative error between the target and actual fertilizer application rates of the applicator generally falls within the range of [6.34%, 7.26%], with an average fertilizer application accuracy of 93.92%, indicating a high level of precision. The experimental results meet expectations, demonstrating that the rice–wheat dual-variable precision fertilizer applicator equipped with the ILMP model performs well and meets operational requirements.

2.5.3 Conclusion

1. By analyzing the optimization principles of the Levy Flight Algorithm (LFA) and Particle Swarm Optimization (PSO), along with the construction mechanism of the Multilayer Perceptron (MLP) neural network model, a nested algorithm framework was designed. Using the programming software Visual Studio Code 1.88.1, the LFA-PSO-MLP model was developed and optimized to predict fertilizer application rate based on dispenser opening degree and rotational speed. Additionally, the Inverse LFA-PSO-MLP model was built to predict the required opening and rotational speed based on the target fertilizer rate.
2. Training results of the MLP, PSO-MLP, and LFA-PSO-MLP models were analyzed. The results show that the R^2 value of the LFA-PSO-MLP model rapidly reached 0.999 during the iteration process, with an average relative error of 1.83%, outperforming the other two models. Validation tests revealed that the average error between the predicted and actual values in the validation set was 2.47%. For the ILMP model, the average relative error in validation tests was 1.82%, indicating high prediction accuracy and good experimental performance.
3. By integrating NDVI (Normalized Difference Vegetation Index) detection technology of rice and wheat canopies with an MLP neural network model, a real-time monitoring and control integrated system for a rice–wheat dual-variable

Table 2.7 Fertilization results of the fertilizer applicator equipped with the ILMP model

Serial number	Target value (g)	Measured value (g)	Measured value per time ($g \cdot min^{-1}$)	Relative error (%)
1	5612.93	5260.72	3507.15	6.70
2	5960.31	5605.16	3803.44	6.34
3	5821.91	5428.03	3752.02	7.26
4	5647.63	5293.44	3528.96	6.69
5	6064.18	5655.98	3770.65	7.22
6	5777.36	5645.73	3763.82	2.33
MAPE (%)				6.08

precision fertilizer applicator was established. Field trials with the LMP model showed a prediction accuracy of 3.49%, while trials with the ILMP model achieved an average accuracy of 93.92%. These results indicate that the prediction model can effectively and accurately determine the required opening and rotational speed parameters for a given target fertilizer amount, enabling efficient and precise variable-rate fertilization in precision agriculture. This contributes to optimal fertilizer resource utilization, environmental protection, and improved ecological and economic benefits.

References

1. Lan S, Liu S (2015) A review of cuckoo search algorithm. Comput Eng Des 36(4):1063–1067
2. Shi YY (2018) Design on variable fertilizer spreader with centrifugal discs in spectral reflectance for rice field. Nanjing Agricultural University, Nanjing
3. Ren L, Tian M, Li J et al (2023) Research status and development analysis of variable rate fertilization technology in China. J Agric Mach Res 45(10):10–15
4. Chen J, Zhao B, Yi S et al (2017) Research status and development strategies of variable rate fertilization technology in China. J Agric Mach Res 39(10):1–6
5. Shi Y, Chen M, Wang X et al (2018) Design and experiment of a centrifugal fertilizer spreader for variable surface fertilization in paddy fields. Trans Chin Soc Agric Mach 49(3):86–93
6. Liang Y, Tang Z, Ji C et al (2023) Structural parameter optimization and experimental analysis of external grooved fertilizer discharger. J Agric Mach Res 45(12):7–14
7. Dun G, Liu W, Du J et al (2022) Optimized design and experimentation of double-helix fertilizer discharger with arc groove. Trans Chin Soc Agric Mach 53(10):118–125
8. Chen M, Shi Y, Wang X et al (2015) Design and experiment of precision fertilizer applicator for wheat based on spectral detection. Trans Chin Soc Agric Mach 46(5):26–32
9. Yuan J, Liu C, Li Y et al (2009) Gaussian processes based bivariate control parameters optimization of variable-rate granular fertilizer applicator. Comput Electron Agric 70(1):33–41
10. Li J, Liu Q, Wang Z (2023) Prediction of soil fertilizer supply based on extreme learning machine model optimized by improved sparrow search algorithm. Chin J Agric Mach 44(10):174–184
11. Zheng H, Liu S, Hou Y et al (2014) Study on the relationship between the ecological balance fertilization model and the fertilizer effect function model. J Agric Resour Environ 31(6):500–505
12. He X (2021) Research on maize yield prediction model based on machine learning. Jilin Agricultural University, Changchun
13. Wang X (2022) Research on soil temperature and moisture prediction technologies based on deep learning. Jilin University, Changchun
14. An S (2006) Research on prediction model of Bursaphelenchus xylophilus based on remote sensing and GIS. Beijing Forestry University, Beijing
15. Jia H, Li Y, Huang S et al (2023) Research on precision fertilization model for maize based on cloud computing and RBF neural network integration. J Maize Sci 31(6):128–134
16. Yang S, Li Y, Shi Y et al (2024) Prediction of spatial distribution of soil organic matter and fertilization plan for strawberry planting zones—a case study of Changfeng County. J Agric Resour Environ:1–11
17. Huang Z, Liu T (2023) Establishment and simulation analysis of precision fertilization model based on neural network algorithm. Mod Agric Res 29(11):45–50
18. Zhang J, Liu G, Ren Z et al (2023) Construction of a fertilizer discharge prediction model for a bivariate fertilization system based on GWO-GRNN. Jiangsu Agric Sci 51(2):210–217

19. Zhang J, Liu G, Zhang D et al (2022) Research status and prospects of control technology for variable spreading of solid granular fertilizer. Jiangsu Agric Sci 50(20):69–77
20. Gong H (2021) Research on real-time prediction and control system for maize midseason tillage variable fertilization based on NDVI. Heilongjiang Bayi Agricultural University, Daqing
21. Shi Y, Chen M, Wang X et al (2015) Design and experiment of the actuator for rice and wheat precision fertilization machine. J South China Agric Univ 36(6):119–124
22. Shi Y, Chen M, Wang X et al (2017) Analysis and experiment of fertilization performance of rice and wheat precision variable fertilization machine. Trans Chin Soc Agric Mach 48(7):97–103
23. Feng X, Li Q, Quan W et al (2021) A review of multi-objective particle swarm optimization algorithms. J Eng Sci Technol 43(6):745–753
24. Ma X, Tan J, Chen S et al (2020) Research on multi-task scheduling optimization particle swarm algorithm in cloud computing. J Electron Meas Instrum 34(8):133–143
25. Wang Q, Guo X (2016) Particle swarm optimization algorithm based on Levy flight. Appl Res Comput 33(9):2588–2591
26. Zhang C, Guo Y, Li M (2021) Review of development and application of artificial neural network model. Comput Eng Appl 57(11):57–69
27. Yan F, Xu J, Li F (2019) Training multi-layer perceptron with chaotic grey wolf optimization algorithm. J Electron Inf Technol 41(4):872–879
28. Ahakonye CAL, Zainudin A, Shanto AJM et al (2024) A multi-MLP prediction for inventory management in manufacturing execution system. Internet Things 26:101156
29. Vaghefi E, Hosseini S, Prorok B et al (2024) Geometrically-informed predictive modeling of melt pool depth in laser powder bed fusion using deep MLP-CNN and metadata integration. J Manuf Process 119:952–963
30. Guo L (2014) Optimal design and experimental research on the type hole of vertical disk soybean seed metering device. Northeast Agricultural University, Harbin
31. Shi YY, Chen M, Wang XC et al (2023) Efficiency analysis and evaluation of centrifugal variable-rate fertilizer spreading based on real-time spectral information on rice. Comput Electron Agric 204:107505

Chapter 3
Design and Modeling of the Centrifugal Variable-Rate Fertilizer Spreader

S. Hu, Xuekai Huang, Yinyan Shi, J. Bo, L. Xu, and L. Lin

Abstract To achieve real-time, site-specific variable-rate fertilization throughout the rice growth cycle and improve both operational efficiency and fertilizer use efficiency, a ground-based dual-disc centrifugal fertilizer applicator equipped with a uniforming cover was developed. The system integrates a near-ground spectral sensing unit, expert decision-making algorithm, and electro-mechanical actuator, and is controlled via a closed-loop feedback system based on an STM32F103 microcontroller to dynamically adjust the fertilizer rate according to crop demand. A three-factor, three-level orthogonal experiment was designed with fertilizer discharge rate, disc rotation speed, and machine forward speed as input variables. The coefficient of variation (Cv) of particle distribution and the relative error of fertilizer rate (γ) were used as performance indicators. Experimental results showed that under the optimal parameter combination ($q = 300$ g/s, $\omega = 600$ r/min, $v = 1.2$ m/s), the system achieved a Cv of 13.82% and a γ of 9.54%. Field trials further validated the system, yielding an average Cv of 14.62% and an average γ of 9.25%, confirming satisfactory uniformity and accuracy. To optimize the spreading performance, a simulation model was established using the Discrete Element Method (DEM). A rotating orthogonal design was conducted to investigate the effects of disc shape, number of vanes, and helical feed angle on spreading uniformity. Results indicated that the disc geometry had the most significant influence on Cv. The optimal configuration—disc curve ratio of 0.276, four vanes, and a feed angle of 47.02°—produced a minimum simulated Cv of 13.29%. The average relative error between simulated and measured values was 8.99%, demonstrating high reliability of the

S. Hu · X. Huang (✉) · J. Bo · L. Xu · L. Lin
College of Engineering, Nanjing Agricultural University, Nanjing, China
e-mail: 2023112024@stu.njau.edu.cn; 2022212011@stu.njau.edu.cn; 2022212013@stu.njau.cn; 2024812087@stu.njau.edu.cn; 2023212009@stu.njau.edu.cn

Y. Shi
College of Engineering, Nanjing Agricultural University, Nanjing, China

Jiangsu Province Engineering Lab for Modern Facility Agriculture Technology & Equipment, Nanjing, China
e-mail: shiyinyan@njau.edu.cn

Y. Shi et al. (eds.), *Technological Progress on Variable Rate Fertilization*, Smart Agriculture 17, https://doi.org/10.1007/978-981-95-4330-4_3

DEM model for performance prediction. In conclusion, the proposed centrifugal variable-rate fertilizer applicator effectively enhances spreading uniformity and rate control accuracy, meeting the demands of precision fertilization in paddy fields and providing a theoretical and technical foundation for upgrading conventional disc-based fertilization systems.

Keywords Centrifugal fertilizer applicator · Variable-rate fertilization · Spreading uniformity · Discrete element method (DEM) · Near-ground spectroscopy · Expert control

3.1 Introduction

Fertilization plays a vital role in rice cultivation, significantly influencing plant growth, biomass accumulation, and final yield. Variable rate fertilization (VRF), which applies nutrients based on spatial variability and crop demand, has emerged as a core technique in precision agriculture [1–3]. However, due to the particularities of paddy fields, most fertilization operations in rice production in China remain manual, resulting in low efficiency, labor intensity, and uneven nutrient distribution. There is an urgent need to enhance the mechanization and intelligence level of rice fertilization.

Centrifugal disc-type spreaders, especially those with large working widths and variable-rate capabilities, have been widely used for surface fertilization in rice fields. These systems can reduce labor input, improve operational efficiency [4], and increase fertilizer use efficiency. In developed countries such as those in Europe and North America, research on centrifugal fertilizer spreaders dates back to the twentieth century [5, 6]. OLIESLAGERS et al. studied the influence of outlet position and disc rotation speed on fertilizer distribution and developed a mathematical model for particle motion, validating it through comparative experiments [7]. CAMPELL et al. proposed a dual-disc spreader with hydraulic flow control and evaluated both open-loop and closed-loop systems for different fertilization scenarios [8]. COETZEE et al. used the Discrete Element Method (DEM) to analyze how parameters such as outlet flow rate and blade angle affect fertilizer distribution uniformity in orchards, demonstrating the strong predictive power of DEM modeling [9, 10].

In recent years, Chinese researchers have made progress in centrifugal spreader development [11]. CHEN Shufa et al. designed a high-clearance self-propelled spreader and evaluated its performance under field conditions [11]. LV Jinqing et al. [1] developed a conical disc-based biased spreading device and optimized its structural parameters via rotational orthogonal experiments to meet potato fertilization requirements. HU Yongguang et al. designed a narrow-row biased centrifugal disc for tea plantations and used EDEM software for simulation and optimization, confirming the reliability of the model through bench testing [12].

Despite these advancements, several challenges remain in the field of centrifugal variable-rate fertilization [13, 14]: (1) fertilizer particle distribution uniformity is difficult to ensure, especially in large-width operations; (2) variable-rate systems often rely on pre-set prescriptions or maps, lacking real-time responsiveness to crop growth dynamics; (3) integrated approaches combining sensor feedback, DEM simulation, and parameter optimization are still underdeveloped [15].

To address these limitations, this study builds upon the research group's previous work on a spectral-based winter wheat fertilization system and proposes a novel dual-disc centrifugal spreader with a fertilizer uniforming cover, designed specifically for surface fertilization in paddy fields [8, 16]. The machine integrates near-ground spectral sensing, expert decision-making algorithms, and a closed-loop actuator system [17]. Structural optimization focuses on key parameters such as disc geometry, vane number, and the inclination angle of the helical feeder.

A DEM-based simulation model is developed using EDEM software to analyze particle movement and distribution during spreading [13, 20]. Design-Expert software is employed to conduct multi-factor simulations and optimize parameters for minimum distribution variation [21–23]. The proposed spreader aims to achieve a single-pass spreading width exceeding 24 m, with a distribution coefficient of variation (Cv) below 15%, thereby ensuring high-efficiency, high-precision, and uniform fertilization in rice cultivation [24].

3.2 Overall Structure and Working Principle

3.2.1 Overall Structure

The overall structure of the sensor-based dual-disc centrifugal fertilizer applicator for surface variable-rate fertilization in rice fields, utilizing spectral technology, is shown in Fig. 3.1. The system primarily consists of a spectral sensor, wireless

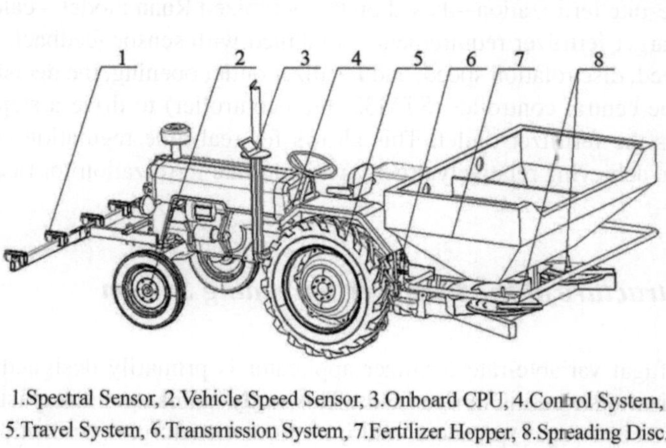

Fig. 3.1 Structural diagram of centrifugal variable-rate fertilizer spreader. 1. Spectral Sensor, 2. Vehicle Speed Sensor, 3. Onboard CPU, 4. Control System, 5. Travel System, 6. Transmission System, 7. Fertilizer Hopper, 8. Spreading Disc

Table 3.1 Technical parameters of fertilizer spreader

Parameter	Value
Overall dimensions (L × W × H) (mm × mm × mm)	2300 × 1450 × 1200
Hitch type	Three-point suspension
Supporting power (kW)	≥65
Fertilizer hopper capacity (L)	1500
Spreading width (m)	≥24
Fertilizer discharge rate ($g{\cdot}s^{-1}$)	100–500
Disc rotation speed ($r{\cdot}min^{-1}$)	300–1000
Operating speed ($m{\cdot}s^{-1}$)	0.8–1.6
Fertilization method	Surface topdressing
Total weight (kg)	640

communication module, vehicle control terminal (CPU), travel system, fertilizer spreading mechanism, and control system. The main technical specifications of the machine are listed in Table 3.1.

The surface variable-rate fertilizer applicator for rice fields is a tractor-mounted implement using a three-point suspension and PTO-driven dual counter-rotating spreading discs. It achieves an effective single-pass spreading width of over 24 m. A closed-loop feedback system independently adjusts the disc rotation speed and fertilizer outlet opening, enabling a high level of automation, uniform distribution, improved operational efficiency, and enhanced fertilizer utilization. This system is well-suited for large-scale, modern agricultural practices in rice-wheat rotation regions.

During operation, the wireless serial communication module transmits real-time NDVI values of the rice canopy, detected by the Greenseeker spectral sensing system, to the onboard control terminal. The embedded expert decision-making system for variable-rate fertilization—based on the optimized Ruan model—calculates the real-time target fertilizer requirement. Combined with sensor feedback on current vehicle speed, disc rotation speed, and fertilizer outlet opening, the decision system instructs the central controller (STM32 microcontroller) to drive a stepper motor that adjusts the fertilizer outlet. This allows for real-time regulation of fertilizer application, achieving relatively precise variable-rate fertilization for rice.

3.2.2 Structure of the Fertilizer Spreading System

The centrifugal variable-rate fertilizer applicator is primarily designed for large-scale fertilizer distribution in rice fields. This study builds upon the existing design of dual-disc centrifugal spreaders, and introduces structural optimizations specifically targeting the spreading system (as shown in Fig. 3.2). The spreading system mainly consists of the frame, fertilizer hopper, centrifugal spreading discs,

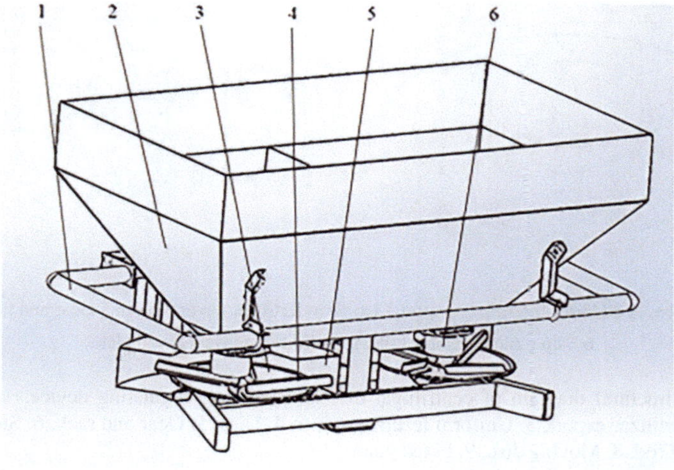

Fig. 3.2 Structural diagram of centrifugal variable fertilizer spreader. 1. rack, 2. particle can, 3. fertilizer disc, 4. fertilizer blades, 5. fertilizer shield, 6. spiral blanking device

spreading vanes, and fertilizer deflector cover, with each component playing a key role in ensuring uniform distribution and operational stability.

During operation, the tractor-towed implement drives the fertilizer spreading system forward, which is mounted on the rear frame via a three-point suspension. Fertilizer stored in the hopper descends under gravity through the discharge outlet and is guided smoothly into the spreading disc by a helical feeding device. As the disc rotates at high speed, the eccentrically positioned vanes propel the fertilizer outward by centrifugal force. The particles are then dispersed through impact and rebound before settling onto the field surface, completing the centrifugal spreading process.

3.3 Structural Design of the Fertilizer Spreading Mechanism

3.3.1 Design of the Spreading Disc

As the core component of the centrifugal fertilizer applicator, the spreading disc assembly primarily consists of a spreading disc, vanes (blades), and a fertilizer uniforming cover, as illustrated in Fig. 3.3a. Based on the overall structural dimensions of the spreader, the diameter of the spreading disc is designed to be 620 mm. To increase the effective spreading width, the disc adopts a conical cross-section with a taper angle α ranging from 0° to 10°.

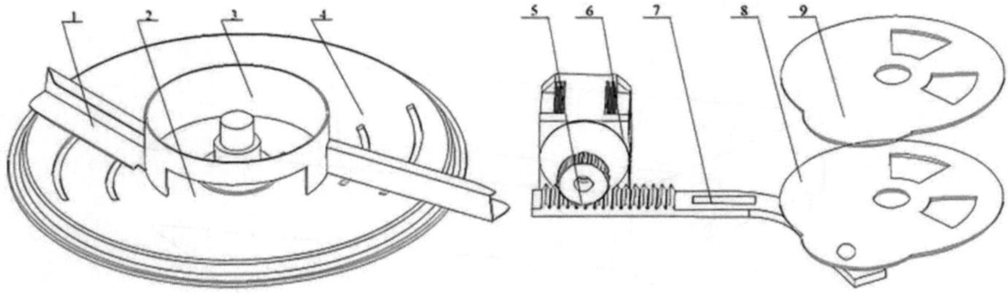

1. Leaves 2 Uniform fertilizer export 3 Uniform fertilizer cover 4. Disc 5 Gear and rack
6 Step motor 7. Connecting rod 8 Moving disc 9 Fixed plate

Fig. 3.3 Structural diagram of centrifugal disc and fertilizer regulating device. 1. Leaves, 2. Uniform fertilizer export, 3. Uniform fertilizer cover, 4. Disc, 5. Gear and rack, 6. Step motor, 7. Connecting rod, 8. Moving disc, 9. Fixed plate

A commonly used dual-blade layout is adopted, with two blades symmetrically arranged 180° apart. The blade inclination angle β, defined as the angle between the blade and the generatrix of the cone, is adjustable via set screws, typically ranging from 6° to 20°.

To match the conical disc and blade configuration, a cylindrical fertilizer uniforming cover is designed with an open-bottom structure for even material distribution. The cover has an outer diameter of 240 mm, height of 80 mm, wall thickness of 5 mm, and an opening height of 45 mm. It is positioned centrally between the two blades. Under centrifugal force, granular fertilizer exits the uniforming cover and is intercepted by the rotating blades in sequence, facilitating uniform spreading across the field.

Due to centrifugal inertia, fertilizer particles are released from the eccentrically mounted blades and follow a projectile trajectory under the combined action of gravity, air resistance, and initial tangential velocity. According to principles of classical mechanics, a necessary condition for particles to be successfully ejected from the disc is that the centrifugal force exceeds the sliding frictional resistance, that is:

$$mr_0\omega^2 > \mu mg \tag{3.1}$$

Where:

m—mass of the fertilizer particle, kg
r_0—distance between the initial position of the particle and the center of rotation,
μ—coefficient of friction between the particle and the spreading disc
ω—angular velocity of the spreading disc, *rad/s*

3.3.2 Design of the Fertilizer Rate Regulation Mechanism

The fertilizer rate regulation mechanism is a critical component of the variable-rate fertilizer applicator. Its structural layout is illustrated in Fig. 3.4b and primarily consists of a stepper motor, gear-rack mechanism, connecting rod, moving and fixed discs, angular displacement sensor, and limit switches.

Based on the target fertilizer rate determined by the lower-level controller, the stepper motor drives the rack-and-pinion linkage to adjust the relative position between the moving disc and the fixed disc, thereby controlling the size of the fertilizer outlet. This adjustment enables dynamic regulation of the discharge flow rate, ranging from 0 to 500 g/s.

The DELIXI LXJM1-8108 limit switch coordinates with the upper-level controller to define the maximum and minimum angular positions of the moving disc, corresponding to the maximum and minimum fertilizer discharge rates. The WDD35D-4 5 k angular displacement sensor measures the current rotational angle of the moving disc, which is then converted into real-time fertilizer flow data and transmitted to the vehicle-mounted CPU for display on the human-machine interface (HMI).

A 57HBP76AL4 stepper motor—chosen for its simplicity, reliability, and precise control—is used to drive the mechanism. It features a rated current of 3 A, an input DC voltage of 10–48 V, a holding torque of 1.5 N·m, and a step angle of 1.8°. The motor is paired with a HYQD40-5742 digital stepper driver to ensure stable and accurate operation.

According to the working principle of the fertilizer application adjustment device, it is simplified as a biased crank slider mechanism during motion analysis, as shown in Fig. 3.4a, and a coordinate system is established. In the figure, e represents eccentricity, mm, r1 represents crank length, mm, r2 represents connecting rod length, mm, l represents slider movement distance, mm, α represents crank angle, °,

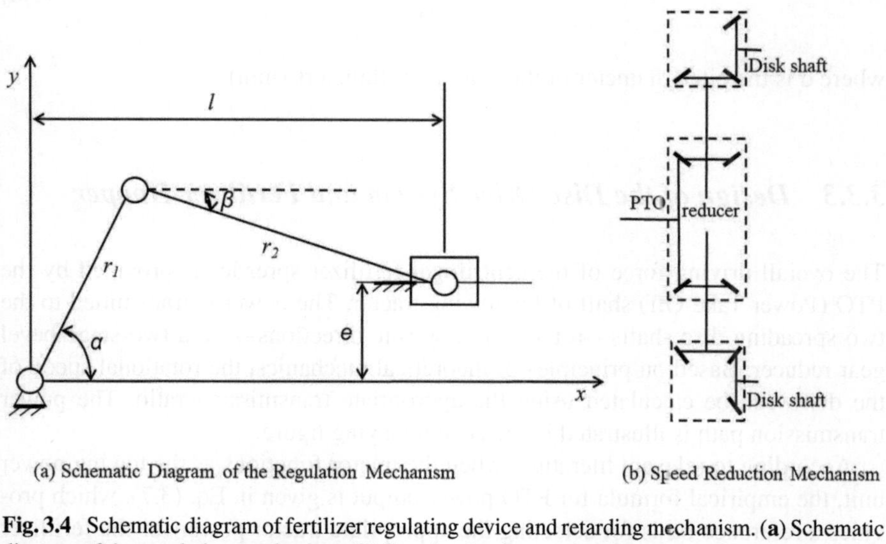

Fig. 3.4 Schematic diagram of fertilizer regulating device and retarding mechanism. (**a**) Schematic diagram of the regulation mechanism. (**b**) Speed reduction mechanism

β represents connecting rod angle, °. According to the motion analysis diagram, the sliding distance l of the slider is:

$$l = r_1 \cos\alpha + r_2 \cos\beta \tag{3.2}$$

Taking the derivative with respect to time *t*, the slider velocity is obtained.

$$v = \frac{dl}{dt} = -\left(r_1 \sin\alpha \frac{d\alpha}{dt} + r_2 \sin\beta \frac{d\beta}{dt} \right) \tag{3.3}$$

From the diagram $r_1 \sin\alpha = r_2 \sin\beta + e$, it follows that, by differentiating with respect to time *t*, we obtain:

$$\frac{d\beta}{dt} = \frac{d\alpha}{dt} \cdot \frac{r_1 \cos\alpha}{r_2 \cos\beta} \tag{3.4}$$

Given the crank angular velocity, substituting Eq. (3.4) into Eq. (3.3) yields:

$$v = \frac{dl}{dt} = -r_1 \omega_1 \frac{\sin(\alpha + \beta)}{\cos\beta} \tag{3.5}$$

Given the known slider velocity v or displacement l, the crank angular velocity $\boldsymbol{\omega}_1$ can be inversely calculated, thereby determining the rotation angle of the moving disc.

According to mechanical principles, in a rack-and-pinion transmission mechanism, the linear velocity of the rack is given by:

$$v = \frac{\pi d n}{60} \tag{3.6}$$

where d is the pitch diameter of the gear, in millimeters (mm).

3.3.3 Design of the Disc Drive System and Fertilizer Hopper

The overall driving force of the centrifugal fertilizer spreader is provided by the PTO (Power Take-Off) shaft of the towing tractor. The power is transmitted to the two spreading disc shafts—rotating in opposite directions—via a two-stage bevel gear reducer. Based on principles of theoretical mechanics, the rotational speed of the discs can be calculated using the appropriate transmission ratio. The power transmission path is illustrated in the accompanying figure.

According to relevant literature, when the tractor functions as the towing power unit, the empirical formula for PTO power output is given in Eq. (3.7), which provides a reference basis for selecting the appropriate tractor power for the fertilizer spreader.

$$P \geq k\left[P_n + \frac{(F_n + mgf)v}{3600}\right] \tag{3.7}$$

Where,

P—*PTO* output power, kW
K—safety factor, generally taken as 1.1–1.2, this article takes 1.2
P_n—average power consumption of the load, kW
F_n—average resistance experienced by the load, N
M—overall quality of the machine, kg
G—gravitational acceleration, m/s^2
F—Friction coefficient of the whole machine, generally taken as 0.6–0.8 in paddy fields, and 0.7 in this article
V—working walking speed, km/h

To reduce the frequency of refilling and improve the operational efficiency of a single fertilization pass, the fertilizer hopper was designed with a total capacity of 1500 L, divided into two independent compartments on the left and right. The hopper adopts an inverted frustum quadrilateral ring geometry to optimize material flow. A filtration mesh is integrated at the inlet to prevent clumped fertilizer particles and foreign impurities from entering the hopper, thereby enhancing the flowability of granular fertilizer and reducing the risk of clogging.

3.4 Control System Design

3.4.1 Spectral Sensing System Design

The sensor-based centrifugal variable-rate fertilizer applicator for rice performs fertilization decisions based on the crop canopy's Normalized Difference Vegetation Index (NDVI). The spectral detection system comprises six GreenSeeker sensors spaced at 0.4 m intervals, one Interface Module for signal coordination, a 12 V power supply, and an adjustable-height mounting frame to maintain a consistent distance of 0.8–1.2 m between the sensors and the rice canopy (as shown in Fig. 3.5).

Operating in an active remote sensing mode, the system uses a red light band (671 ± 6 nm) and near-infrared band (780 ± 6 nm) to capture canopy reflectance data. The spectral signals are converted into electrical signals and transmitted via a CAN bus to the interface module. The data are then remotely transferred to the onboard CPU through the E61-TTL-1 W wireless serial communication module for fertilization decision-making.

As the foundational step in the fertilization process, the accuracy and representativeness of spectral data acquisition are of critical importance. Sampling frequency plays a key role: a higher sampling rate increases data accuracy and spatial representativeness but also imposes greater demands on system hardware and

Fig. 3.5 Picture of sampling test for spectrum data

Table 3.2 Relationship between sampling frequency and distribution uniformity coefficient, coefficient of variation

Sampling frequency f (Hz)	Parameter	Test plot								
		Artificial fertilizer spreading			Traditional machine spreading			Variable machine spreading		
0.8	C_u	92.34	90.86	93.64	93.72	92.16	91.75	94.26	91.87	92.43
	C_v	15.75	16.34	14.29	14.76	13.64	15.07	10.75	12.43	14.82
1.2	C_u	91.84	91.31	90.67	92.46	91.72	93.53	93.07	93.72	91.67
	C_v	14.38	15.76	14.25	14.53	13.74	14.26	13.25	14.03	15.07
1.6	C_u	90.48	91.52	90.03	91.54	90.96	91.03	92.41	91.72	92.64
	C_v	14.06	13.87	14.35	13.27	12.46	14.82	12.08	11.56	13.72
2.0	C_u	89.46	90.72	88.37	91.53	90.46	91.72	92.48	90.37	91.05
	C_v	11.52	12.87	14.69	12.75	13.09	12.91	10.85	12.06	11.75
2.4	C_u	87.51	86.72	88.91	89.77	90.35	89.71	90.28	91.06	89.43
	C_v	10.84	12.67	11.73	11.64	12.81	10.92	11.89	10.75	14.86
2.8	C_u	89.72	85.49	84.61	89.78	86.34	88.96	91.72	90.41	86.75
	C_v	7.68	9.72	11.56	10.48	12.68	9.75	6.67	10.81	9.76

processing; a lower rate facilitates faster processing but may fail to reflect spatial variability in crop growth.

The GreenSeeker RT200 sensor operates within a sampling frequency range of 0.7–3.33 Hz. To balance data accuracy (evaluated via the coefficient of variation of NDVI values) and system performance, the Christiansen Uniformity Coefficient (CU) is used to assess the influence of sampling frequency on NDVI variability and determine the optimal rate. The sampling test setup is shown in Fig. 3.5, and the results are listed in Table 3.2. The Christiansen Uniformity Coefficient is calculated using the following equation:

$$C_u = \left(1 - \frac{\sum_{i=1}^{n} \left| f_i - \overline{f} \right|}{\sum_{i=1}^{n} f_i}\right) \times 100\% \tag{3.8}$$

Random sampling experiments were conducted in three treatment zones: (1) manual broadcasting, (2) uniform fertilization by a conventional spreader, and (3) variable-rate fertilization based on demand using the developed machine. As shown in Table 3.2, the highest uniformity coefficient was 94.26% (variable-rate zone), and the lowest was 84.61% (manual zone). The maximum coefficient of variation (Cv) was 16.34% (manual zone), while the minimum was 6.67% (variable-rate zone). These results indicate that while the sampling frequency fff has little effect on the uniformity coefficient, it significantly affects the coefficient of variation.

Based on the analysis of the data in the table, to ensure a uniformity coefficient ≥90% and a coefficient of variation ≤15% in all three treatment zones, the optimal system sampling frequency was determined to be f = 1.6 Hz.

3.4.2 Hardware Design of the Control System

As the core component of the centrifugal variable-rate fertilizer applicator, the control system consists of both hardware and software. It handles tasks including sensor data acquisition, communication, data storage, real-time control of actuators, and display via the human-machine interface (HMI).

The hardware framework of the control system is illustrated in Fig. 3.6, and it mainly comprises: mainly including power supply, voltage regulation module, GPS module, sensor module, wireless communication module, core controller module, driver module, vehicle control terminal, human-computer interaction module, etc.

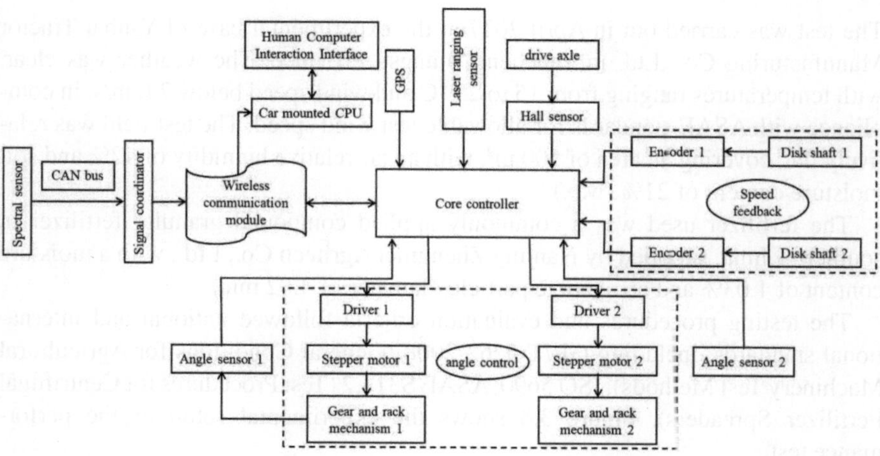

Fig. 3.6 Block diagram of control system for variable-rate fertilizer spreader

The power system employs a 24 V 36 Ah lithium battery to provide independent power supply for the entire control system, ensuring stable operation unaffected by external factors. A voltage regulation module distributes power to different electrical components. The core controller is based on the STM32F103 microcontroller, which receives spectral data transmitted via the E61-TTL-1 W wireless communication module and performs fertilization decision-making.

The vehicle-mounted control terminal, operating on a Windows OS, calculates the real-time nitrogen demand of rice crops based on NDVI input. Combined with feedback from a Hall sensor (detecting tractor speed), rotary encoders (monitoring the rotational speed of the dual centrifugal discs), and an angular displacement sensor (monitoring the fertilizer outlet opening), the STM32 generates PWM signals to control the motor driver module, thereby adjusting the outlet aperture of the fertilizer hopper and regulating the fertilizer application rate.

For real-time positioning, the system utilizes the ATK-NEO-6 M-V23 GPS module, which also assists in vehicle speed estimation. The human-machine interface (HMI) is implemented using a WAT-T8060-104 multifunctional configuration terminal based on an ARM platform, featuring a 10.4-inch TFT LCD screen for visualizing collected data, storing field operation information, and transmitting operational commands (Fig. 3.6).

3.5 Performance Test

To evaluate the fertilizer distribution uniformity and application accuracy of the variable-rate centrifugal fertilizer spreader, a performance test was conducted.

3.5.1 Test Conditions

The test was carried out in April 2017 at the experimental base of Yanhai Tractor Manufacturing Co., Ltd. in Yancheng, Jiangsu Province. The weather was clear, with temperatures ranging from 15 to 25 °C and wind speed below 2.0 m/s, in compliance with ASAE standards for allowable test wind speed. The test field was relatively flat, covering an area of 500 m^2, with an air relative humidity of 42% and soil moisture content of 21% (w.b.).

The fertilizer used was a commonly applied compound granular fertilizer in southern China, supplied by Nanjing Zhengmei Agritech Co., Ltd., with a moisture content of 1.03% and an average particle diameter of 4.02 mm.

The testing procedures and evaluation criteria followed national and international standards, including GB/T 5262-2008 (General Conditions for Agricultural Machinery Test Methods), ISO 5690, ASAE S314.2 (Test Procedures for Centrifugal Fertilizer Spreaders). Figure 3.8 shows the experimental setup of the performance test.

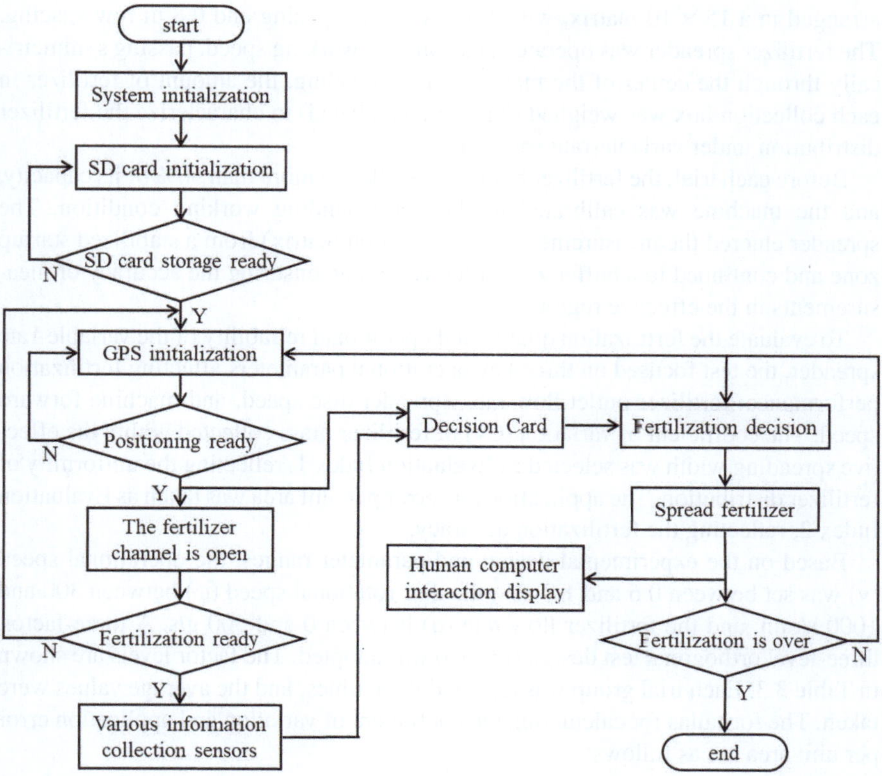

Fig. 3.7 Flow-chart of control system

Fig. 3.8 Performance test of spreading fertilizer

3.5.2 Test Method

During fertilizer application, plastic film was laid across the test area (30 m × 14 m) to minimize fertilizer loss. To reduce measurement errors caused by particle rebound, soft cotton fabric was attached to the inner walls of the fertilizer collection boxes. A total of 150 boxes (dimensions: 50.8 cm × 40.6 cm × 10.2 cm) were arranged in a 15 × 10 matrix, with 1.5 m column spacing and 0.8 m row spacing. The fertilizer spreader was operated at a constant working speed, passing symmetrically through the center of the matrix. After spreading, the amount of fertilizer in each collection box was weighed (bagged and labeled) to characterize the fertilizer distribution under variable-rate spreading.

Before each trial, the fertilizer hopper was filled to more than 50% of its capacity, and the machine was calibrated to the corresponding working condition. The spreader entered the measurement area (collection matrix) from a stabilized startup zone and continued to a buffer zone after spreading, ensuring the accuracy of measurements in the effective region.

To evaluate the fertilization quality and operational reliability of the variable-rate spreader, the test focused on three key operational parameters affecting fertilization performance: fertilizer outlet flow rate, spreader disc speed, and machine forward speed. The coefficient of variation (Cv) of fertilizer mass collected within the effective spreading width was selected as Evaluation Index 1, reflecting the uniformity of fertilizer distribution. The application rate error per unit area was taken as Evaluation Index 2, reflecting the fertilization accuracy.

Based on the experimental design and parameter ranges, the operational speed (v) was set between 0.6 and 1.8 m/s, the disc rotational speed (ω) between 300 and 1000 r/min, and the fertilizer flow rate (q) between 0 and 500 g/s. A three-factor, three-level orthogonal test design ($L9(3^4)$) was adopted. The factor levels are shown in Table 3.3. Each trial group was repeated three times, and the average values were taken. The formulas for calculating the coefficient of variation and application error per unit area are as follows:

$$\gamma = \frac{|M / S - Q|}{Q} \times 100\% \tag{3.9}$$

In the formula

γ—Error in fertilizer application per unit area, %
M—Total mass of fertilizer particles in the experimental area, g
S—experimental area area, m^2
Q—Theoretical target fertilization amount, g

$$C_V = \frac{S_D}{\overline{X}} \times 100\% \tag{3.10}$$

Table 3.3 Factors and levels of orthogonal test

Level	Fertilizer discharge flow rate q (g/s)	Disc rotational speed ω (r/min)	Forward speed v (m/s)
1	150	400	0.8
2	300	600	1.2
3	450	800	1.6

Where: $S_D = \left[\sum_{i=1}^{n}\left(X_i - \bar{X}\right)^2 / (n-1)\right]^{\frac{1}{2}}$, $\bar{X} = \frac{1}{n}\sum_{i=1}^{n} X_i$.

In the formula

C_V—coefficient of variation of particle distribution, %
S_D—The quality standard deviation of fertilizer granules, g
$\bar{X}$—The average quality of fertilizer granules, g
X_i—The mass of fertilizer granules in the first collection box, g
n—Quantity of collection boxes

3.5.3 Test Results and Analysis

The orthogonal performance test scheme and corresponding results are shown in Table 3.4, where A, B, and C represent the level values of fertilizer outlet flow rate (q), disc rotational speed (ω), and forward speed (v), respectively.

Range analysis of Table 3.4 reveals that the significance of influencing factors varies across different evaluation indices. When prioritizing the coefficient of variation of particle distribution (Cv) as the primary performance metric, the order of influence is A (fertilizer outlet flow rate) > B (disc speed) > C (forward speed). The optimal level combination is A1B3C2. Cv increases with increasing A, decreases with increasing B, and exhibits a non-monotonic trend—first decreasing then increasing—with C.

Table 3.4 Results of orthogonal test

Test no.		Test factors			Coefficient of variation Cv (%)	Fertilizer application error γ (%)
		A	*B*	*C*		
1		1	1	1	17.73	14.66
2		1	2	3	15.51	10.83
3		1	3	2	12.67	12.47
4		2	1	2	15.19	10.78
5		2	2	1	16.72	8.46
6		2	3	3	15.57	11.21
7		3	1	3	19.92	12.54
8		3	2	2	18.44	10.13
9		3	3	1	17.08	13.29
Coefficient of variation (cv)	k_1	15.30	17.61	17.18	*A* > *B* > *C*	
	k_2	15.83	16.89	15.43		
	k_3	18.48	15.11	17.00		
	R	3.18	2.50	1.75		
Fertilizer application error (γ)	k_1	12.65	12.66	12.14	*B* > *A* > *C*	
	k_2	10.15	9.81	11.13		
	k_3	11.99	12.32	11.52		
	R	2.50	2.85	1.01		

When prioritizing the fertilizer application error (γ) as the evaluation metric, the order of influence becomes B > A > C, with the optimal level combination being A2B2C2. The application error γ demonstrates a consistent trend of initial decrease followed by increase with the rise of each factor (A, B, and C).

These findings indicate that the variable-rate centrifugal spreader exhibits different sensitivity patterns depending on the performance criterion, and the optimal working parameters should be selected accordingly to achieve either higher distribution uniformity or greater fertilization accuracy.

The results of the variance analysis are shown in Table 3.5. For the coefficient of variation Cv, the factor significance follows the order FA > FB > FC, indicating that Factor A (fertilizer outlet flow rate) has the most significant effect on Cv, followed by Factor B (disc rotational speed), and then Factor C (forward speed), with $P < 0.05$. For the fertilizer application error γ, the order is FB > FA > FC, showing that Factor B has the most significant impact on γ, followed by Factor A, with Factor C having the least effect, also with $P < 0.05$. These findings are consistent with the range analysis. Combining both range and variance analyses, the optimal factor level combinations vary depending on the evaluation priority: when prioritizing distribution uniformity (Cv), the optimal combination is A1B3C2, yielding Cv = 12.67% and γ = 12.47%; when prioritizing application accuracy (γ), the optimal combination is A2B2C2, which was not included in the original orthogonal design and was thus tested separately, resulting in Cv = 13.82% and γ = 9.54%. Comparing the best-performing combinations under different evaluation priorities and referencing relevant standards for fertilizer spreader performance [2] (Cv $\leq$ 15% and $\gamma \leq$ 10%), the combination A2B2C2 is selected as the optimal configuration. This corresponds to a fertilizer outlet flow rate q = 300 g/s, disc rotational speed ω = 600 r/min, and forward speed v = 1.2 m/s, under which the spreader achieves the best overall performance

Table 3.5 Analysis of variance

Evaluation index	Source of variation	Sum of squares	Degrees of freedom	Mean square	F	Significance
C_v	Adjusted model	32.92	6	5.49	7.13	
	A	17.41	2	8.70	6.55	**
	B	9.99	2	4.99	3.76	*
	C	5.53	2	2.76	2.08	*
γ	Adjusted model	26.23	6	4.37	8.15	
	A	10.08	2	5.04	9.40	**
	B	14.59	2	7.29	13.59	**
	C	1.55	2	0.78	1.45	*

* indicates significant difference, ** indicates extremely significant difference

3.5.4 Field Experiment

To verify the accuracy of the optimal factor level combination derived from the aforementioned orthogonal performance tests and to enhance the fertilizer spreading performance of the variable-rate centrifugal fertilizer spreader, a field experiment was conducted in June 2017 at the rice experimental base of Jinmansui Agricultural Development Co., Ltd., Dongtai, Jiangsu Province. The weather conditions were clear, with temperatures ranging from 25 to 36 °C and wind speeds below 2.0 m/s. The total test area covered 1.3 ha, with a relative humidity of 31%. The experimental procedure followed the standard method described in the study. A photograph of the field test setup is shown in Fig. 3.9.

During the experiment, the operational parameters of the fertilizer spreader were adjusted to the optimal level combination: fertilizer outlet flow rate q = 300 g/s, disc rotational speed ω = 600 r/min, and forward speed v = 1.2 m/s. Six groups of repeated trials were conducted (across six test plots), and the results are shown in Table 3.6. The results indicate that the field coefficient of variation (Cv) of fertilizer particle distribution ranged from a maximum of 15.71% to a minimum of 12.13%, with an average of 13.97%, compared to a mean value of 9.19% from the bench performance tests. The fertilization error (γ) ranged from 10.76% to 8.24%, with an

Fig. 3.9 Picture of field validation test

Table 3.6 Results of field test

Test no.	Coefficient of variation (Cv) (%) Estimated value (%)	Measured value (%)	Relative error (%)	Fertilizer application error (γ) (%) Estimated value (%)	Measured value (%)	Relative error (%)
1	13.82	15.16	9.70	9.54	9.95	4.29
2		14.84	7.38		8.62	9.64
3		12.13	12.23		9.03	5.35
4		13.07	5.43		10.76	12.79
5		15.71	13.67		8.24	13.63
6		12.89	6.73		10.18	9.78
Mean		13.97	9.19		9.46	9.25

average of 9.46%, which closely aligns with the mean value of 9.25% from the performance tests. Both evaluation metrics meet the operational standards for fertilizer application machinery, demonstrating that the designed centrifugal variable-rate fertilizer spreader significantly improves both distribution uniformity and fertilization accuracy. This provides a theoretical foundation for enhancing traditional empirically based fertilization practices with centrifugal spreaders.

3.6 Conclusion

1. Based on the uniformity and accuracy requirements of variable top dressing operation of rice, a kind of sensor based double-disk centrifugal uniform fertilizer cover type rice surface variable fertilizing machine was designed, and a supporting variable fertilizing operation control system was developed, and the corresponding fertilizer discharge performance test was carried out to achieve efficient, accurate and uniform variable fertilizing of rice.
2. With STM32F103 as the core controller, the sensor data information is integrated, the feedback information is coordinated, and the fertilizing strategy model is combined to respond to the target fertilizer demand. The stepping motor is guided by the decision system to control the opening of fertilizer discharge, and the real-time fertilizing amount of crops is adjusted online, so as to realize the accurate variable fertilizing of rice in a relative sense.
3. The fertilizing performance test results show that the fertilizer discharge flow A, disc rotating speed B and travel speed C of the spreader have different effects on different evaluation indexes. Taking the variation coefficient of particle distribution Cv as the evaluation index, the primary and secondary influencing factors are A, B and C, and the optimal combination of factors is A1B3C2; The relative error γ of fertilizer amount is the evaluation index, the primary and secondary factors are B, A and C, and the optimal combination of factor level is A1B3C2; Comparing different optimal combination schemes, select the relatively better factor level combination A2B2C2, i.e. $q = 300$ g/s, $\omega = 600$ r/min, $v = 1.2$ m/s, $Cv = 13.82\%$, $\gamma = 9.54\%$. The fertilizing performance of the whole fertilizing machine is the best.
4. The field test results show that the average variation coefficient Cv is 13.97%, and the average error with the performance test is 9.19% under the optimal combination of fertilizer application parameters; The average error γ of fertilizer application amount is 9.46%, and the average error with performance test is 9.25%. The results show that the designed centrifugal variable fertilizer spreader can improve the uniformity of fertilizer distribution and the accuracy of fertilizer amount, and provide a technical support for the traditional empirical fertilization of centrifugal fertilizer spreader.

References

1. Song H (2016) Optimization design and simulation of quantitative screw conveyor. Qingdao University of Science and Technology, Qingdao
2. Wang Y (2021) Design and experiment study of side throwing centrifugal fertilizer spreader for lotus root field. Huazhong Agricultural University, Wuhan
3. Yu F, Shi WM (2015) Nitrogen use efficiencies of major grain crops in China in recent 10 years. Acta Pedol Sin 52(6):1311–1324
4. Shi YY (2018) Design on variable fertilizer spreader with centrifugal discs in spectral reflectance for rice field. Nanjing Agricultural University, Nanjing
5. Shi YY, Chen M, Wang XC et al (2018) Design and experiment of variable-rate fertilizer spreader with centrifugal distribution cover for rice paddy surface fertilization. Trans Chin Soc Agric Mach 49(3):86–93+113
6. Chen M, Shi YY, Wang XC et al (2015) Design and experiment of variable rate fertilizer applicator based on crop canopy spectral reflectance. Trans Chin Soc Agric Mach 46(5):26–32
7. Olieslagers R, Ramon H, Baerdemaeker JD (1996) Calculation of fertilizer distribution patterns from a spinning disc spreader by means of a simulation model. J Agric Eng Res 63(2):137–152
8. Campbell CM, Fulton JP, McDonald TP et al (2010) Spinner-disc technology to enhance the application of poultry litter. Appl Eng Agric 26(5):759–767
9. Liedekerke PV, Tijskens E, Dintwa E et al (2009) DEM simulations of the particle flow on a centrifugal fertilizer spreader. Powder Technol 190(3):348–360
10. Liedekerke PV, Tijskens E, Ramon H et al (2009) Discrete element simulations of the influence of fertiliser physical properties on the spread pattern from spinning disc spreaders. Biosyst Eng 102(4):392–405
11. Chen SF, Zhang SP, Sun XZ et al (2012) Design and experiment of self-propelled high-ground-clearance spreader for paddy variable-rate fertilization. Trans Chin Soc Agric Eng 28(11):16–21
12. Hu YG, Yang YC, Xiao HR et al (2016) Simulation and parameter optimization of centrifugal fertilizer spreader for tea plants. Trans Chin Soc Agric Mach 47(5):77–82
13. Villette S, Cointault F, Piron E et al (2005) Centrifugal spreading: an analytical model for the motion of fertiliser particles on a spinning disc. Biosyst Eng 92(2):157–164
14. Fulton JP, Hancock DW, Higgins SF et al (2005) Distribution pattern variability of granular VRT applicators. Trans ASAE 48(6):2053–2064
15. Tijskens E, Ramon H (2003) Discrete element modeling for process simulation in agriculture. J Sound Vib 266(3):493–514
16. Fulton JP, Shearer SA, Chabra G et al (2001) Performance assessment and model development of a variable-rate, spinner-disc fertilizer applicator. Trans ASAE 44(5):1071–1081
17. Xu L (2018) The brachistochrone curve experiment. Educ Teach Forum (4):190–191
18. Yang YC (2016) Operating parameter optimization and experiment on centrifugal fertilizer distributor for tea fields. Jiangsu University, Zhenjiang
19. Meng XZ, Shi JH, Wang XL et al (2015) Experimental research on influential factor for scattering performance of disc manure spreader. J Anhui Agric Sci 43(27):335–337
20. Guo LF (2014) Optimization design and experiment study on hole of vertical disc soybean seed-metering device. Northeast Agricultural University, Harbin
21. Jin LL, Ji CY, Fang HM et al (2015) Numerical simulation of mixing process of fertilizer particles in continuous mixer of variable rate fertilizer applicator. Acta Agriculturae Zhejiangensis 27(2):261–265

22. Malone KF, Xu BH (2008) Determination of contact parameters for discrete element method simulations of granular systems. Particuology 6:520–530
23. Liu GH, Liu W, Liu Z et al (2010) Predicting build-up rate of drilling tool by orthogonal regression method with four factors and three levels. J Oil Gas Technol 32(3):108–112+409
24. ASAE S341.2 (1999) Procedure for measuring distribution uniformity and calibrating granular broadcast spreaders

Chapter 4
Mechanical Properties of Fertilizer Particles and Crop Straw Under Discrete Element Modeling

J. Wang, Lei Wang, Yinyan Shi, C. Xia, Xiaochan Wang, and Z. Zhao

Abstract In modern agricultural mechanization, granular materials such as fertilizers and crop residues (e.g., straw) are widely involved in operations like fertilizing, seeding, harvesting, and mulching. The mechanical behavior of these particles—such as movement, stress response, breakage, and deposition—directly affects application uniformity, operation efficiency, and mechanical wear. The Discrete Element Method (DEM), as a core tool for analyzing multiparticle systems, has become increasingly valuable in the structural optimization and performance prediction of agricultural machinery. This chapter focuses on two representative granular subjects—fertilizer granules and individual crop straw stems—and systematically investigates their physical and mechanical properties, contact parameter calibration, and DEM modeling strategies. The content includes the construction of a comprehensive granular property database, simulation-based analysis of fertilizer discharge dynamics, and a mechanical model of straw stems incorporating failure characteristics. These studies provide both theoretical foundations and numerical references for intelligent control and equipment design in precision agriculture.

Keywords Fertilizer pellet · Physical and mechanical properties · Measurement method · The database · Compound fertilizer

---

J. Wang · C. Xia
College of Engineering, Nanjing Agricultural University, Nanjing, China
e-mail: wangjihao@stu.njau.edu.cn; 9213010210@stu.njau.edu.cn

L. Wang (✉) · Y. Shi · X. Wang
College of Engineering, Nanjing Agricultural University, Nanjing, China

Jiangsu Province Engineering Lab for Modern Facility Agriculture Technology & Equipment, Nanjing, China
e-mail: leiwang@njau.edu.cn; shiyinyan@njau.edu.cn; wangxiaochan@njau.edu.cn

Z. Zhao
College of Information and Electrical Engineering, China Agricultural University, Beijing, China
e-mail: zhaozhangcau@cau.edu.cn

Y. Shi et al. (eds.), *Technological Progress on Variable Rate Fertilization*, Smart Agriculture 17, https://doi.org/10.1007/978-981-95-4330-4_4

4.1 Introduction

Traditional extensive fertilization practices often lead to excessive fertilizer application, causing not only resource inefficiency but also risks of chemical accumulation in crops and environmental degradation. In recent years, under the guidance of the national "Double Reduction" policy, increasing emphasis has been placed on adopting scientific and rational fertilization strategies to enhance fertilizer use efficiency. As a core component of precision agriculture, mechanized variable-rate fertilization provides a technical pathway for demand-oriented and accurate fertilizer application, playing a crucial role in promoting sustainable agricultural modernization. This technology has attracted significant attention from researchers aiming to optimize fertilization and nutrient management, improve fertilizer utilization, ensure food safety, and foster sustainable agricultural development. The movement of fertilizer particles is largely governed by intrinsic properties (e.g., density, moisture content, diameter, Poisson' s ratio, and elastic modulus) and contact parameters (e.g., restitution coefficient, static friction coefficient, and rolling friction coefficient). Among these, the friction coefficients mainly regulate particle motion, whereas the restitution coefficient characterizes particle stability [1, 2]. In this study, the intrinsic properties of compound fertilizer particles were measured through physical experiments, while the contact parameters that are difficult to determine directly were calibrated using discrete element simulations.

The discrete element method (DEM) has been continuously refined since its introduction by Gundall et al. and has been widely applied in the study of agricultural materials. Hao et al. developed a DEM model for yam interacting with sandy loam, calibrating both the mechanical properties of yam and the contact parameters of the substrate. Zhang et al. used spherical particles of varying radii to fill the contour of rice seeds, identifying the particle size that optimized simulation accuracy and computational efficiency. Wang et al. constructed sunflower seed models using different numbers of spherical particles and obtained simulated angles of repose and packing densities that closely matched experimental observations. Lu et al. compared ellipsoidal and multi-sphere modeling approaches, showing that ellipsoidal modeling could accurately reproduce the motion of wheat seeds, offering a refined DEM approach for wheat seed simulations. To reduce computational demands, the particle size in DEM models can be appropriately increased, allowing simulations to be completed within reasonable time frames. Ren et al. applied similarity theory and dimensional analysis to scale up pulverized iron coal particles, calibrate the contact parameters for enlarged particles, and validate the scaling method's accuracy. Zou et al. similarly calibrated contact parameters for quicklime powder using particle scaling theory, achieving simulation results in close agreement with physical experiments [3].

In the past, numerous scholars employed statistical and tracer methods to investigate the interaction between straw and agricultural machinery. These approaches are not only time-consuming and labor-intensive but also primarily capture the macroscopic motion of straw, making it difficult to analyze the micro-stress behavior of straw. In recent years, with advances in computational technology, discrete element method (DEM)-based simulations have been widely adopted to study the interaction between wheat straw and machinery, facilitating the design and optimization of agricultural equipment and enhancing operational performance. For instance, DEM has been used to develop a straw-soil-tillage machine interaction model [4], allowing the analysis of stress, motion, and spatial distribution of straw and soil particles, as well as the forces exerted on machinery during tillage. The flexible straw model combined with CFD-DEM coupling has been applied to simulate the rice-wheat separation process in a combine harvester [5]. Shi et al. established a simulation model for the working process of straw shredding and no-till planter operation, verified the accuracy and effectiveness of the DEM model through field experiments, and determined the optimal operating parameters for the no-till planter. Zhou et al. conducted both DEM simulations and field experiments for three cultivation modes—traditional rotary tiller, straw rotary burying and returning machine, and subsoiling plus straw rotary burying and returning machine—and demonstrated that the subsoiling plus straw rotary burying and returning approach achieves the most uniform vertical and horizontal distribution of straw in the soil.

4.2 Materials and Methods

4.2.1 Materials

In the present study, Huachang 15-15-15 compound fertilizer was adopted as the experimental material, with a particle density of 1585 kg/m^3. Stainless steel, the primary material used for the spreader's spreading components, was also employed for the cylinders and plates that come into direct contact with fertilizer particles. The experimental setup and materials are depicted in Fig. 4.1.

The physical and mechanical properties of straw are affected by multiple factors, including crop cultivar, cultivation environment, irrigation schedule, climatic

Fig. 4.1 Test materials. **(a)** Compound fertilizer. **(b)** Cylinder

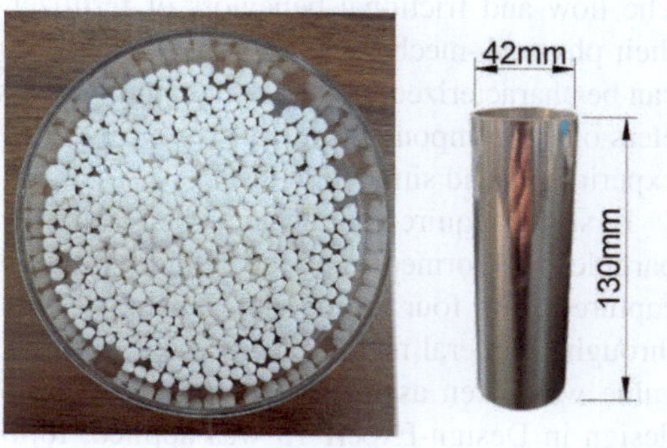

(a) compound fertilizer (b) cylinder

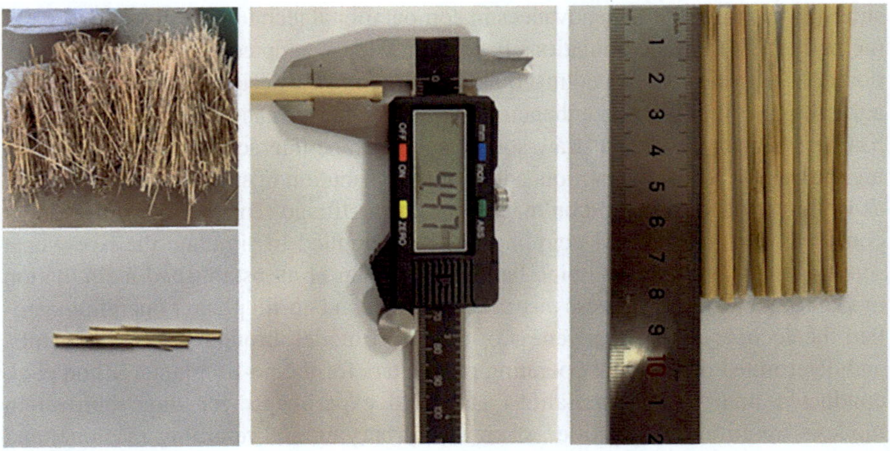

Fig. 4.2 Sample preparation and measurement

conditions, fertilization methods, chemical treatments, plant health status, and the position of straw within the wheat plant. Among these factors, plant maturity and moisture content are particularly critical. The straw samples used in this study were collected from the experimental field at the Sihong Demonstration Base, Sihong County, Jiangsu Province, China. The wheat variety tested was Qian Mai 088, with a plant height ranging from 1000 to 1200 mm and an average of 5 internodes. Samples were cut from the third internode above the root; each sample had a length of 80 mm, an average diameter of 4.22 mm, a wall thickness of 0.43 mm, and a moisture content of 13.73% (wet basis, w.b.). As shown in Fig. 4.2, four groups of samples were prepared (ten specimens per group) to assess the tensile, compressive, bending, and shear properties of wheat straw.

4.2.2 Methods

The flow and frictional behaviors of fertilizer particles—closely associated with their physical–mechanical properties and the characteristics of contact materials—can be characterized by the resting angle. In this study, the discrete element parameters of the compound fertilizer were calibrated through a combination of physical experiments and simulation tests.

First, the required parameters were acquired via experiments. A pile of fertilizer particles was formed using the cylinder lifting method, and images of the pile were captured from four orthogonal directions. The resting angle was then determined through unilateral measurement using MATLAB image processing, and the mean value was taken as the actual resting angle. Subsequently, the Plackett–Burman design in Design-Expert 13 was applied, followed by simulation tests in EDEM 2021.2 to measure the resting angle after the system stabilized. The analysis of variance (ANOVA) results from the Plackett–Burman design was used to identify significant factors influencing the resting angle. A steepest ascent experiment was further conducted to narrow the parameter range; thereafter, the Box–Behnken design and corresponding ANOVA were employed to establish the regression model and regression equation describing the relationship between the resting angle and the significant parameters. Nonsignificant terms were excluded to optimize the model. By solving the regression equation with the actual resting angle set as the target value, the optimal parameter combination was determined. Finally, validation simulations were performed under these optimized conditions to compare the simulated and measured resting angles, thereby verifying the calibration accuracy.

For straw testing, an electronic universal testing machine (CMT6104, Meister Industrial System Co., Ltd., China) was used for tensile tests (Fig. 4.3a), while a texture analyzer (TMS-PRO, Food Technology Corporation, USA) was employed for compressive, bending, and shear tests (Fig. 4.3b–d). From these experiments, the tensile strength, compressive strength, elastic modulus, and shear strength of the straw were obtained.

During the tensile tests, both ends of each straw sample were clamped with a 15 mm grip length, the loading speed was set to 10 mm/min, and the sampling frequency was 10 Hz (Fig. 4.3a). The tensile strength was calculated using Eq. (4.1):

$$\sigma_t^{\max} = \frac{F_t^{\max}}{A} \tag{4.1}$$

Where, $F_t^{\max}$ is the maximum tensile load, N; A is the cross-sectional area of test straw, m^2.

Radial compression tests were conducted at the same loading speed and sampling frequency (Fig. 4.3b), and the compressive strength was determined by Eq. (4.2):

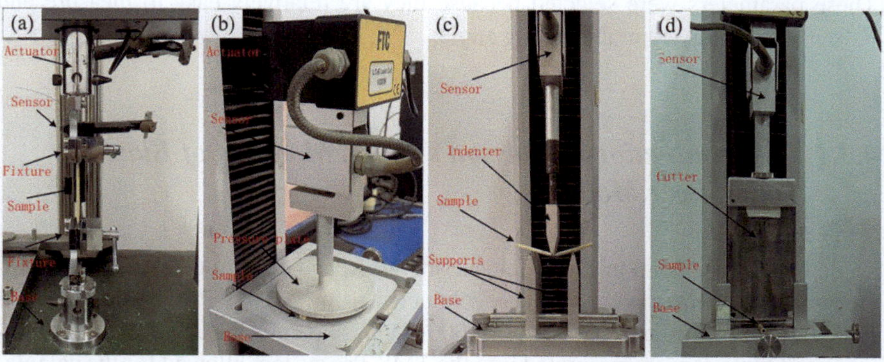

(a) Tensile test (b) Compression test (c) Three-point bending test (d) Shear test

Fig. 4.3 Straw mechanical properties test. (**a**) Tensile test. (**b**) Compression test. (**c**) Three-point bending test. (**d**) Shear test

$$\sigma_p^{\max} = \frac{F_p^{\max}(D-t)}{Lt^2} \tag{4.2}$$

Where, $F_p^{\max}$ is the maximum pressure load, N; D is the outer diameter, mm; t is the wall thickness, mm; L is the length of test straw, mm.

The elastic modulus was derived from a three-point bending test (Fig. 4.3c), in which straw samples were supported at both ends on steel fixtures and loaded at 10 mm/min with a cylindrical indenter (diameter: 8 mm). The elastic modulus was then calculated according to Eq. (4.3):

$$E_b = \frac{F_b L_s^3}{48SI} \tag{4.3}$$

Where, F_b is the pressure load provided for the cylinder indenter, N; L_s is the distance between two supports (40 mm); S is the deflection (deformation), mm; I is the inertia moment of the cross-section because the section of straw can be approximately a ring, it can be written as the inertia moment:

$$I = \frac{\pi}{64}\left[D^4 - (D-2t)^4\right] \tag{4.4}$$

Where, D is the outer diameter, mm; t is the wall thickness, mm.

4.3 Determination of Actual Angle of Repose

The intrinsic parameters of the compound fertilizer were obtained through physical testing, with a focus on the collision recovery coefficient, static friction coefficient, and rolling friction coefficient—all of which were used for simulation calibration. These parameters were selected because they are relatively easy to measure and have minimal impact on the resting angle.

4.3.1 Determination of Poisson's Ratio, Modulus of Elasticity and Shear Modulus

To determine the Poisson's ratio, elastic modulus, and shear modulus of the fertilizer granules, an extrusion test was conducted using a TMS-Pro texture analyzer (FTC, USA; maximum test force of 1000 N), as shown in Fig. 4.4. The displacement and force sensors integrated into the TMS-Pro system allowed for the calculation of the granules' Poisson's ratio, elastic modulus, and shear modulus based on the following equations:

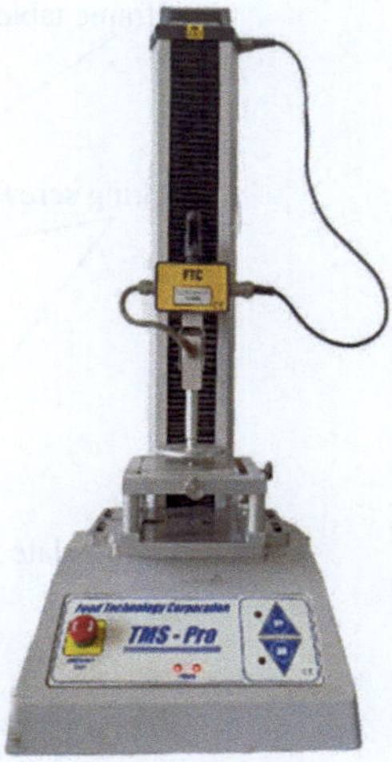

Fig. 4.4 TMS-Pro texture analyzer

$$\begin{cases} v = \left| \dfrac{W_1 - W_2}{H_1 - H_2} \right| \\ E = \dfrac{FH_1}{A(H_1 - H_2)} \\ G = \dfrac{E}{2(1+v)} \end{cases} \tag{4.5}$$

In these equations, v, E, and G represent Poisson's ratio, elastic modulus, and shear modulus, respectively; W_1 and W_2 denote the width of the granules before and after loading; H_1 and H_2 indicate the thickness before and after loading; F is the applied load; and A refers to the cross-sectional area of the granules.

A total of 30 intact fertilizer particles were randomly selected for the extrusion test using the TMS-Pro texture analyzer. The average results showed that the Poisson's ratio, elastic modulus, and shear modulus of the compound fertilizer were 0.24, 2.85×10^7, and 1.15×10^7 Pa, respectively.

4.3.2 Determination of Collision Recovery Coefficient

Based on the kinematic theory of the inclined plate collision test (Fig. 4.5), the collision recovery coefficient of the compound fertilizer particles was derived.

Fertilizer granules were released from a fixed height H_0, impacted an inclined plate set at angle θ, and subsequently underwent oblique projectile motion before landing on the receiving plate. To evaluate the collision recovery coefficient, the distance between the inclined plate and receiving plate was varied, and the test was

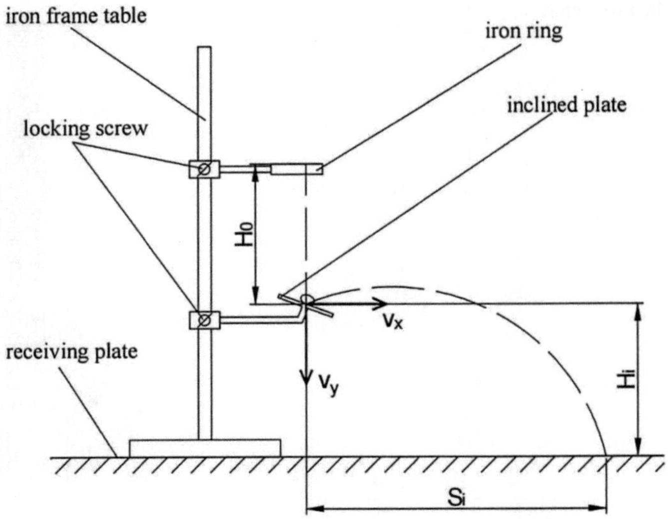

Fig. 4.5 Schematic diagram of inclined plate collision test

repeated twice to record the horizontal displacements (S_1, S_2) and vertical displacements (H_1, H_2) of the particles. The inclined plate could be fitted with different collision materials (e.g., stainless steel plate or fertilizer particle bed), enabling measurement of the collision recovery coefficient under various contact conditions. The kinematic equation was applied to estimate the approach velocity of fertilizer particles as they struck the inclined plate:

$$v_0 = \sqrt{2gH_0} \tag{4.6}$$

When the particles followed an oblique projectile trajectory, Eq. (4.7) was obtained according to the principle of oblique motion.

$$\begin{cases} S_i = v_x t \\ H_i = v_y t + \dfrac{1}{2} g t^2 \end{cases} (i = 1{,}2) \tag{4.7}$$

where S_1 and H_i ($i = 1, 2$) represent the horizontal and vertical displacements of the fertilizer particles, respectively. From Eq. (4.8), the horizontal and vertical velocity components after collision with the inclined plate were derived.

$$\begin{cases} v_x = \sqrt{\dfrac{gS_1S_2\left(S_1 - S_2\right)}{2\left(H_1S_2 - H_2S_1\right)}} \\ v_y = \dfrac{H_1 v_x}{S_1} - \dfrac{gS_1}{2v_x} \end{cases} \tag{4.8}$$

The collision recovery coefficient is defined as the ratio of post-collision separation velocity to pre-collision approach velocity, and is calculated as follows:

$$e = \left|\frac{v_n'}{v_n}\right| = \left|\frac{\sqrt{v_x^2 + v_y^2}\sin\left(\theta + \arctan\frac{v_z}{v_z}\right)}{v_0\cos\theta}\right| \tag{4.9}$$

where v_n' is the post-collision separation velocity, v_n is the pre-collision approach velocity, and θ is the inclination angle of the steel plate. Based on repeated experiments, the collision recovery coefficients between compound fertilizer particles and the steel plate were found to range from 0.05–0.47 to 0.04–0.40, respectively.

4.3.3 Determination of Friction Coefficient

4.3.3.1 Static Friction Coefficient

The static friction coefficient was determined via the slant method. Taking the fertilizer–steel pair as an example, fertilizer particles were first fixed with adhesive tape to prevent rolling before they slid on the inclined surface [2]. The test procedure was as follows: particles adhered to the tape were placed on the steel incline, the inclination angle was gradually increased, and the angle at which sliding initiated was recorded (Fig. 4.6).

To mitigate the impact of particle irregularity on the fertilizer–fertilizer static friction coefficient, relatively larger and flatter particles were selected and affixed to form a granular bed. The same procedure was repeated, and a schematic of the device is provided in the subsequent figure. Based on the measured inclination

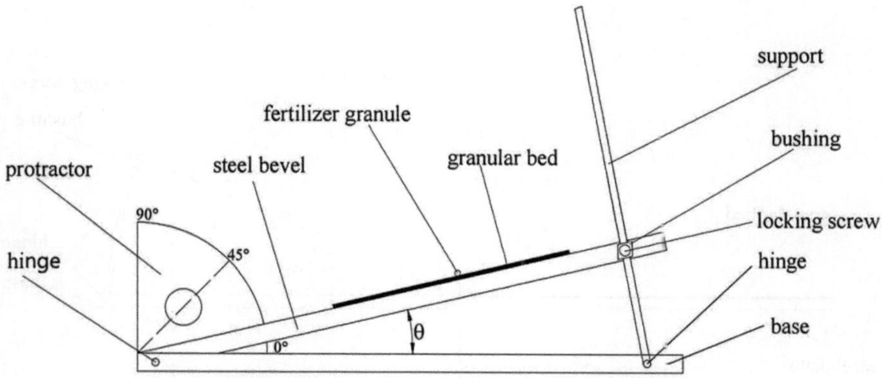

Fig. 4.6 Schematic diagram of measuring device for static friction coefficient of fertilizer-fertilizer

angle, the static friction coefficients for the fertilizer–steel and fertilizer–fertilizer pairs were calculated using Eq. (4.10).

$$\mu = \tan\theta \tag{4.10}$$

where μ is the static friction coefficient and θ is the slope angle at which sliding began. Multiple trials showed that the fertilizer–fertilizer and fertilizer–steel static friction coefficients ranged between 0.62–0.93 and 0.27–0.49, respectively.

4.3.3.2 Rolling Friction Coefficient

According to the results from Sect. 4.3.1, a steel plate inclined at 23° was chosen to ensure fertilizer particles could roll freely along its surface. During testing, individual pellets were released from rest at the top of the slope and allowed to roll across a horizontal steel plate, with the rolling distance measured using a tape. Additionally, particles were affixed to the steel plate to create a flat particle bed, and the same procedure was repeated to evaluate the rolling friction coefficient of the compound fertilizer. A schematic of the measurement device is shown in Fig. 4.7.

Based on the principle of energy conservation, the rolling friction coefficients for fertilizer–fertilizer and fertilizer–steel contacts were calculated via Eq. (4.11). Repeated trials indicated that the rolling friction coefficients for fertilizer–fertilizer and fertilizer–steel pairs ranged from 0.39–0.91 to 0.10–0.22, respectively.

$$\begin{cases} mgH = \mu_1 mg\left(H\cot\alpha + S_1\right) \\ mgH = \mu_1 mgH\cot\alpha + \mu_2 mgS_2 \end{cases} \tag{4.11}$$

In the equation, H denotes the height from the release point to the bottom of the slope, α is the slope angle, μ_1 and μ_2 correspond to the fertilizer–fertilizer and fertilizer–steel rolling friction coefficients, while S_1 and S_2 represent the rolling distances across the horizontal steel plate and the particle bed, respectively.

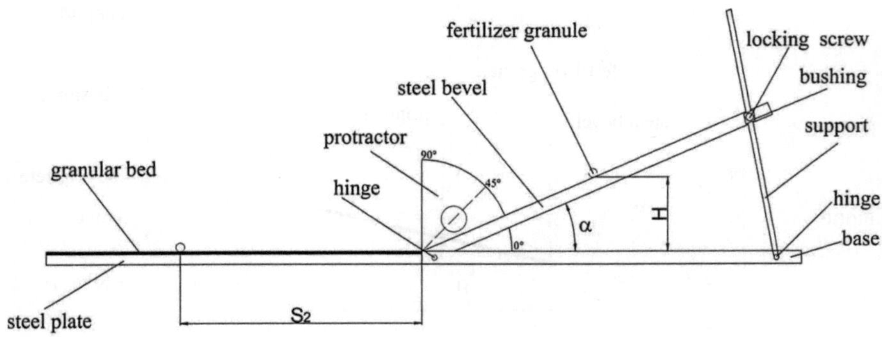

Fig. 4.7 Schematic diagram of measuring device for rolling friction coefficient of fertilizer-fertilizer

4.3.4 Simulation Model

The selection of a suitable contact model for fertilizer particles mainly depends on particle size, moisture content, and surface adhesion [6, 7]. Previous studies have shown that adhesive forces can be neglected when the moisture content of clay does not induce plastic deformation [8–10]. In this study, the moisture content of the fertilizer was measured using an electronic balance (accuracy: 0.01 g) and an electric blast drier (maximum temperature: 255 °C). The measured moisture content was 5.6%, indicating that surface adhesion was negligible. Therefore, the motion of compound fertilizer particles was considered to result solely from minor elastic deformation, and the Hertz–Mindlin (no-slip) contact model was selected to describe their mechanical interactions.

To determine particle size, 200 fertilizer granules were randomly sampled, and their length, width, and thickness were measured using Vernier calipers (accuracy: 0.02 mm). These three dimensions corresponded to the X, Y, and Z axes of the particle model. The equivalent diameter was calculated using Eq. (4.12) [11, 12].

$$D = \sqrt[3]{LWH} \tag{4.12}$$

where D is the equivalent diameter of the compound fertilizer particles, and L, W, H, are the length, width and thickness of the compound fertilizer particles, respectively.

Further, the sphericity of the compound fertilizer particles can be derived from Eq. (4.13) [13, 14].

$$\phi = \frac{D}{L} \tag{4.13}$$

Experimental measurements showed that the average equivalent diameter of the compound fertilizer particles was 3.55 mm, with a mean sphericity of 90.9%, a particle size range of 2.52–5.17 mm, and a frequency distribution illustrated in Fig. 4.8. Given that the average sphericity exceeded 85%, the particles were classified as spherical, allowing each particle to be directly represented by a single sphere in the model [3, 4]. Figure 4.9b shows the construction of the discrete element model using a particle diameter of 3.55 mm. Based on the particle size distribution obtained from physical measurements, particles were generated in the simulation following a random distribution scheme, with the minimum and maximum size limits set to 0.7 and 1.45 times the average diameter, respectively [15, 16].

By reviewing relevant literature [5], the intrinsic properties of stainless steel were identified. Table 4.1 summarizes the intrinsic and contact characteristics of the compound fertilizer and stainless steel [17].

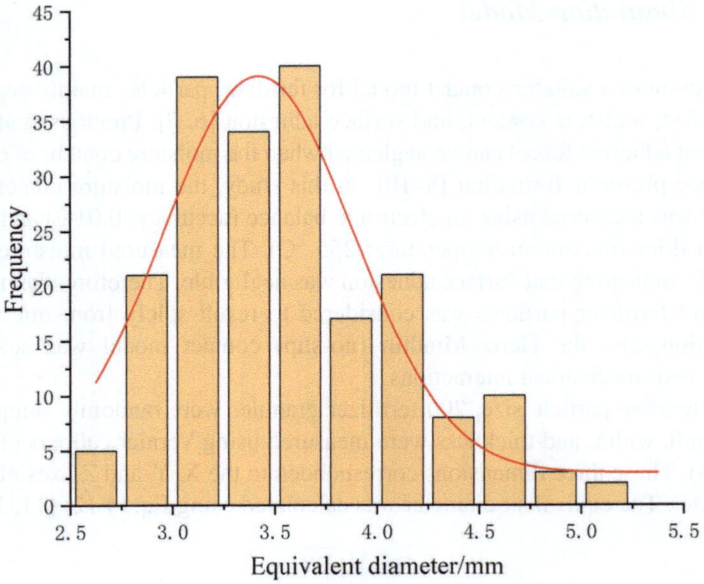

Fig. 4.8 Particle size distribution of compound fertilizer

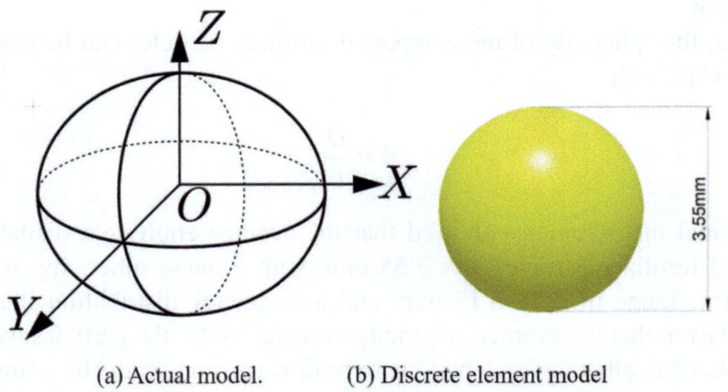

Fig. 4.9 Actual model and discrete element model of compound fertilizer particles. (**a**) Actual model. (**b**) Discrete element model

4.4 Discrete Element Modeling of Straw

The commercial discrete element method (DEM) software EDEM (Version 2018, Altair, www.altair.com) was used to establish a discrete element model of a single wheat straw [18–20]. The modeling process consisted of four steps: (1) Pretreatment: defining the calculation field, as well as the size and arrangement of particles; (2) Defining the physical properties of particles (e.g., mass, motion laws, particle relationships, and failure criteria) and applying boundary conditions; (3) Calculating

Table 4.1 Simulation parameters of discrete element

Materials	Parameters	Numerical value
Fertilizer	Density (kg/m^3)	1585
	Poisson's ratio	0.24
	Shear modulus (Pa)	1.15 × 10^7
Steel	Density (kg/m^3)	7800
	Poisson's ratio	0.3
	Shear modulus (Pa)	7 × 10^{10}
Fertilizer-fertilizer	Collision recovery coefficient	0.05–0.47
	Static friction coefficient	0.62–0.93
	Rolling friction coefficient	0.39–0.91
Fertilizer-steel	Collision recovery coefficient	0.04–0.4
	Static friction coefficient	0.27–0.49
	Rolling friction coefficient	0.1–0.22

the force, velocity, acceleration, and position of the particles; (4) Processing the simulated particle images.

4.4.1 Particle Distribution

Measurements of actual straw revealed a thin-walled tubular structure [21–23]. To accurately replicate its mechanical behavior while ensuring computational feasibility, the position of each sphere in the straw model was defined using the application programming interface (API) of EDEM software. A total of 3696 spherical particles (each with a diameter of 0.52 mm) were assembled into a thin-walled tube model with an inner diameter of 3.5 mm, an outer diameter of 4.5 mm, and a length of 80 mm (Fig. 4.10). The straw model was organized into 154 radially arranged circular layers, with each layer containing 24 spherical particles.

4.4.2 Contact Force Between Particles

The Hertz–Mindlin (no-slip) contact model used in the straw model is the default model in EDEM, offering high accuracy and computational efficiency. In this model, the normal force component is derived from Hertzian contact theory, while the tangential force is based on the formulations proposed. Both normal and tangential forces include damping components, with the damping coefficient related to the coefficient of restitution [24]. Tangential friction follows Coulomb's friction law, whereas rolling friction is modeled using a torque approach—this approach is contact-independent and maintains a constant direction.

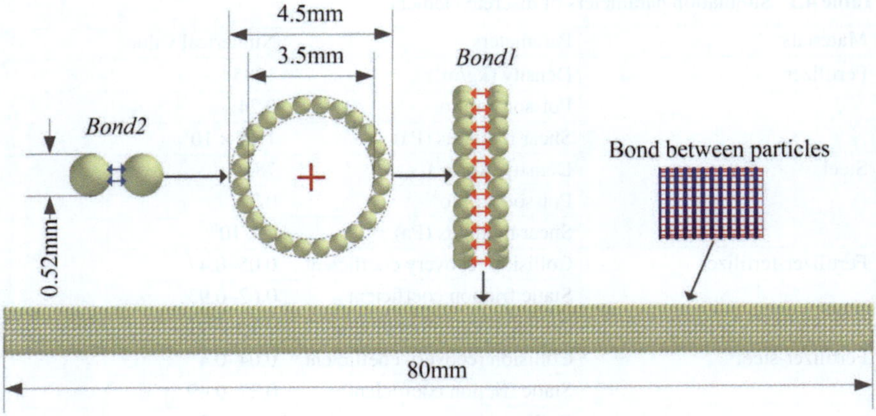

Fig. 4.10 Schematic diagram of the straw model

Table 4.2 Spring stiffness and damping coefficients used in the Hertz-Mindlin (no-slip) contact model

	Normal direction	Tangential direction
Spring stiffness constant (*K*)	$K_n = \frac{4}{3}E^*\sqrt{R^*\delta_n}$	$K_t = 8G^*\sqrt{R^*\delta_n}$
Damping coefficient (*C*)	$C_n = 2\sqrt{\frac{5}{6}}\beta\sqrt{S_n m^*}\overline{v_n^{rel}}$	$C_t = 2\sqrt{\frac{5}{6}}\beta\sqrt{K_t m^*}\overline{v_t^{rel}}$
$S_n = 2E^*\sqrt{R^*\delta_n}, \beta = \frac{\ln e}{\sqrt{\ln^2 e + \pi^2}}$		

For the Hertz-Mindlin (no-slip) model, the total normal force F_n is the sum of the elastic normal force and the damping normal force component:

$$F_n = -K_n\delta_n + C_n\overline{v_n^{rel}} \tag{4.14}$$

Where, K_n and C_n are defined in Table 4.2. δ_n is the normal overlap and $\overline{v_n^{rel}}$ is the normal component of the relative velocity, which provides a spring repulsion force to push the colliding particles apart and buffer part of the relative kinetic energy.

In the tangential direction, the force-displacement relationship is influenced by both normal and tangential loading conditions. The relative tangential velocity v_t^{rel} generated during particle interaction represents the incremental spring energy storage, corresponding to the tangential elastic deformation at the contact surface. The damper dissipates energy through tangential motion, simulating the tangential plastic deformation of the contact particles. The total tangential force F_t is constrained by Coulomb friction law:

$$F_t = \min\left\{\mu_s F_n, K_t\delta_t + C_t\overline{v_t^{rel}}\right\} \tag{4.15}$$

Where, K_t and C_t are defined in Table 4.2. δ_t is the tangential overlap and μ_s is the static friction coefficient. In the equation of Table 4.2, R^*, E^*, G^*, m^* and e are equivalent radii, Young's modulus, shear modulus, mass and recovery coefficient, respectively.

4.4.3 Contact Parameters Between Particles

In the present study, the particle density of straw was derived from the previously determined particle count and corresponding volume, with the actual mass of straw used for calculation; this yielded a density of 460 kg/m^3. The coefficients of restitution, static friction, and rolling friction—both between straw particles and between straw particles and steel—were determined by following methodologies reported in previous studies [17, 25].

When utilizing EDEM software, the selection of an appropriate time step is critical, with two primary types involved [26]: the Rayleigh Time Step and the Fixed Time Step. The Rayleigh Time Step represents the optimal time step in EDEM and is applied throughout the entire straw model to compute the transmission of shear, compressive, and support forces within the mold. This time step is automatically calculated by the software to ensure stable simulations within a reasonable computational duration. In contrast, the Fixed Time Step refers to the constant interval between consecutive calculations, which is influenced by factors such as model complexity (including the spatial arrangement and size of straw particles) and available computational power. Typically, the Fixed Time Step is set to a value between 2% and 40% of the Rayleigh Time Step. In this study, considering particle size, spatial distribution, bonding characteristics, and computational constraints, the Fixed Time Step was set to 2.5×10^{-8} s—equivalent to 2% of the Rayleigh Time Step. The fundamental parameters of the simulation model are summarized in Table 4.3.

4.4.4 Bond Method Between Particles

A laser confocal microscope (Model OLS-4100, Olympus Corporation, Japan) was used to observe and analyze the fiber structure on the straw surface (Fig. 4.11). The results showed that the fibers are well-defined, thin, relatively uniform in length, and densely arranged. Previous studies have indicated that supporting fibers are primarily aligned along the axial direction of straw growth, while numerous small transverse fibers radially connect these axial fibers. As a result, the mechanical properties of straw differ significantly between the axial and radial directions. Therefore, the particles in the straw model were bonded using two distinct bond parameters (Bond1 and Bond2) for the axial and radial directions, respectively

Table 4.3 Material physical and contact mechanical properties parameters

Item	Simulation parameters	Values
Material parameters	Density of steel (kg·m^{-3})	7800
	Density of straw (kg·m^{-3})	460
	Poisson's ratio of steel	0.3
	Poisson's ratio of straw	0.4
	Shear modulus of steel (GPa)	10
	Shear modulus of straw (GPa)	0.001
Interaction parameters	Straw-straw restitution coefficient	0.28
	Straw-steel restitution coefficient	0.2
	Straw-straw static friction coefficient	0.54
	Straw-steel static friction coefficient	0.8
	Straw-straw rolling friction coefficient	0.05
	Straw-steel rolling friction coefficient	0.01
Particle diameter (mm)		0.52

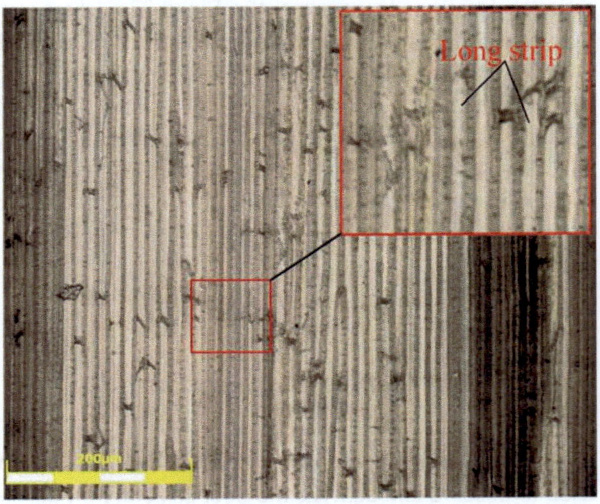

Fig. 4.11 Micro morphology of the straw surface

(Fig. 4.10). Relative displacement between particles induces bond deformation, and the overall deformation of the straw is the cumulative deformation of all bonds. Given the nonlinear mechanical behavior of biomaterials, a nonlinear contact model combined with a linear bond model was adopted in this study, due to limitations of the software.

The bonded-particle model (BPM) was employed to bond straw particles. Developed by Potyondy and Cundall, BPM was initially used to simulate granite strain under biaxial and triaxial compression and is effective for modeling fracture in brittle materials. It has also been applied to simplified equipment operations, such as ball mills, straw kneading or crushing machines, and grape stemming machines.

In BPM, two adjacent particles are bonded at their contact points, with the bond functioning as a distributed spring array on the particle cross-section. The mechanical properties of the bond are described by the contact force and moment at the interface. Particle translation and rotation, caused by contact forces, external forces, and adhesive forces, are determined using Newton's second law; meanwhile, contact forces resulting from relative particle motion are governed by the force-displacement law. The forces and moments acting on a single bond are illustrated in Fig. 4.12 and can be calculated using Eq. (4.16):

$$\begin{cases} \Delta F_n^b = -k_n^b \cdot A^b \cdot v_n \cdot \Delta t \\ \Delta F_s^b = -k_s^b \cdot A^b \cdot v_t \cdot \Delta t \\ \Delta M_n^b = -k_s^b \cdot J \cdot \omega_n \cdot \Delta t \\ \Delta M_s^b = -k_n^b \cdot \dfrac{J}{2} \cdot \omega_t \cdot \Delta t \end{cases} \tag{4.16}$$

Where

$$\begin{cases} A^b = \pi \cdot \left(R^b\right)^2 \\ J = \dfrac{1}{2} \pi \cdot \left(R^b\right)^4 \end{cases} \tag{4.17}$$

Here, k_n^b and k_s^b represent the normal and tangential bond stiffness, v_n and v_t denote the normal and tangential velocity, ω_n and ω_t are the normal and tangential angular velocity, and Δt corresponds to the Rayleigh time step. J and A^b indicate the polar moment of inertia and the cross-sectional area of the bond, respectively. R^b is the bond radius, which is constrained by the smallest particle diameter among the

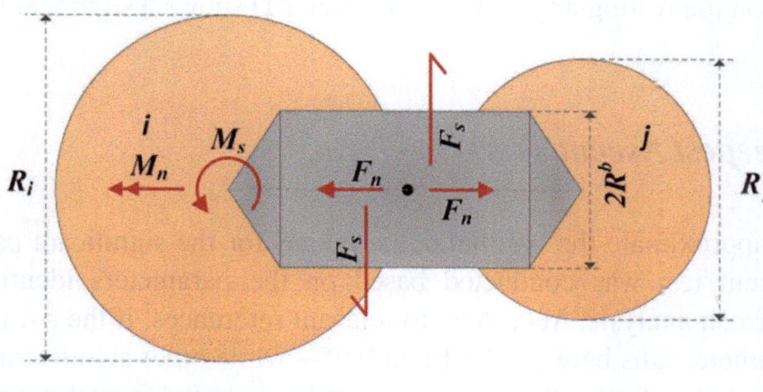

Fig. 4.12 Schematic diagram of bonded-particle model (BPM)

contacting particles. The forces F^b and torque M^b acting in the normal (F_n and M_n) and tangential (F_s and M_s) directions vary according to the relative normal and tangential displacements as well as the normal and tangential rotations of the particles.

By specifying the critical normal stress σ_c and critical tangential stress τ_c, the maximum stress leading to bond fracture is defined, establishing the ultimate stress criteria for the bond:

$$\begin{aligned} \sigma &= \frac{F^b_{n,Total}}{A^b} + \frac{2M^b_s}{J} R^b \\ \tau &= \frac{F^b_{s,Total}}{A^b} + \frac{M^b_n}{J} R^b \end{aligned} \tag{4.18}$$

When σ is greater than σ_c or τ is greater than τ_c, the bond fracture disappears. When the bond between the two particles breaks, the particles can only transmit force, but cannot transmit the torque. The contact action is constrained by Hertz-Mindlin (no-slip) contact model.

4.5 Results

4.5.1 Plackett-Burman Test

Some simulation parameters have no significant impact on the resting angle and thus cannot be calibrated using the resting angle alone [8, 27]. To improve the accuracy of simulation calibration [28], contact parameters were screened to identify those with significant effects on the resting angle, using Design-Expert 13 (Stat-Ease Inc.). The parameter levels used in the simulation tests are presented in Table 4.4.

The design and results of the Plackett-Burman test are summarized in Table 4.5. Analysis of variance (ANOVA) was performed on the Plackett-Burman results to evaluate the effect and significance of six parameters on the resting angle, as shown in Table 4.6. According to the results, Parameters A, C, and E exert highly significant effects on the resting angle, while Parameter D shows a significant influence.

4.5.2 Steepest Ascent Test

To quickly approximate the optimal value ranges for the significant parameters, a steepest ascent test was conducted based on the parameters identified by the Plackett-Burman analysis. According to relevant references, if the p-value of a significant parameter falls between 0.01 and 0.05—while other significant parameters have p-values below 0.01—that parameter can be excluded from the steepest ascent

Table 4.4 Parameter levels of Plackett-Burman test

Parameters	Low level	High level
A (fertilizer-fertilizer collision recovery coefficient)	0.05	0.47
B (fertilizer-fertilizer static friction coefficient)	0.62	0.93
C (fertilizer-fertilizer rolling friction coefficient)	0.39	0.91
D (fertilizer-steel collision recovery coefficient)	0.04	0.4
E (fertilizer-steel static friction coefficient)	0.27	0.49
F (fertilizer-steel rolling friction coefficient)	0.1	0.22

Table 4.5 Protocol and results of Plackett-Burman test

No.	Test factors A	B	C	D	E	F	Resting angle (°)
1	−1	1	1	1	−1	−1	14.9
2	−1	−1	−1	−1	−1	−1	16.12
3	1	−1	1	1	1	−1	18.83
4	−1	−1	1	−1	1	1	14.09
5	1	1	−1	−1	−1	1	19.46
6	1	1	−1	1	1	1	25.25
7	1	1	1	−1	−1	−1	17.35
8	−1	1	−1	1	1	−1	19.63
9	−1	1	1	−1	1	1	15.56
10	1	−1	−1	−1	1	−1	21.61
11	−1	−1	−1	1	−1	1	17.38
12	1	−1	1	1	−1	1	16.55

test. In this study, although Parameter D is significant, its effect is considerably weaker than that of Parameters A, C, and E. Consequently, only Parameters A, C, and E were included in the steepest ascent test. These parameters were varied from low to high values in specific increments, while the remaining parameters were set at intermediate levels.

As shown in Table 4.7, increasing Parameters A, C, and E initially increased the relative error between the simulated and actual resting angles, followed by a subsequent decrease; meanwhile, the simulated resting angle of the compound fertilizer

Table 4.6 ANOVA of Plackett-Burman test results

Source of variance	Mean square	F-value	*p*-value
Models	17.36	20.77	0.0022
A	38.06	45.54	0.0011
B	4.78	5.71	0.0623
C	40.96	49.02	0.0009
D	5.81	6.95	0.0461
E	14.54	17.40	0.0087
F	0.0019	0.0022	0.9641

Note: *p*-value <0.05 indicates significant difference; *p*-value <0.01 indicates highly significant difference, the same below

Table 4.7 Scheme and results of steepest ascent test

No.	factors			Resting angle (°)	Relative error (%)
	A	C	E		
1	0.05	0.35	0.25	15.43	22.67
2	0.15	0.45	0.30	17.82	10.67
3	0.25	0.55	0.35	18.71	6.23
4	0.35	0.65	0.40	19.06	4.48
5	0.45	0.75	0.45	21.04	5.46
6	0.55	0.85	0.50	22.28	11.69

increased continuously. Therefore, the optimal interval is likely near the value corresponding to Test No. 4.

4.5.3 Box-Behnken Test

Based on Table 4.7, the parameter levels corresponding to Test No. 4 were designated as the center point for the Box-Behnken test, with the levels of Test No. 3 and Test No. 5 serving as the low and high points, respectively. This selection was justified by the steepest ascent results, which indicated that the resting angle at Test No. 4 exhibited the smallest relative error compared to the actual resting angle. Table 4.8 presents the test procedure and results, including a total of 17 trials with 5 center-point replications.

The analysis of variance (ANOVA) results for the quadratic regression model are shown in Table 4.8. As presented in Table 4.9, Parameters A, C, E, AC, A^2, and C^2 exert highly significant effects on the resting angle—confirming the model's validity. The p-value for the quadratic regression model is less than 0.0001, while the p-value for the lack-of-fit term is 0.0842 (greater than 0.05), indicating a satisfactory model fit. The regression model demonstrates high reliability, with a coefficient of determination (R^2) of 0.9830, an adjusted R^2 of 0.9612, a coefficient of variation (CV) of 1.45%, and an Adeq Precision value of 27.5977—all of which reflect a high

Table 4.8 Scheme and results of Box-Behnken test

No.	Test factors			Resting angle (°)
	A	C	E	
1	1	0	1	21.06
2	−1	0	−1	16.54
3	0	0	0	19.36
4	0	1	−1	19.39
5	0	0	0	19.7
6	0	0	0	19.43
7	1	1	0	20.91
8	1	−1	0	18.78
9	0	−1	1	20.66
10	0	0	0	19.72
11	−1	0	1	19.75
12	0	1	1	22.78
13	0	−1	−1	17.8
14	−1	1	0	18.52
15	0	0	0	19.39
16	1	0	−1	17.9
17	−1	−1	0	18.78

Table 4.9 ANOVA of Box-Behnken test quadratic model

Source	Sum of squares	df	F-value	*p*-value
Model	32.31	9	45.06	<0.0001
A	3.2	1	40.16	0.0004
C	3.89	1	48.84	0.0002
E	19.91	1	249.83	<0.0001
AC	1.43	1	17.92	0.0039
AE	0.0006	1	0.0078	0.9319
CE	0.0702	1	0.8813	0.3791
A^2	2.75	1	34.56	0.0006
C^2	1.21	1	15.19	0.0059
E^2	0.0432	1	0.5417	0.4857
Residual	0.5578	7		
Lack of fit	0.4348	3	4.71	0.0842
Pure error	0.123	4		
Cor total	32.87	16		

degree of model accuracy. Based on the above analysis, the second-order regression equation describing the relationship between the angle of repose and the significant parameters is as follows:

$$\begin{aligned}\alpha &= 19.52 + 0.6325A + 0.6975C + 1.5775E + 0.5975AC \\ &-0.0125AE + 0.1325CE - 0.80875A^2 + 0.53625C^2 + 0.10125E^2\end{aligned} \tag{4.19}$$

In the regression model, Parameters AE, CE, and E^2 had p-values greater than 0.05, meaning they exerted no significant effect on the resting angle. These terms were therefore removed to optimize the regression model and enhance its reliability and accuracy. Table 4.10 presents the ANOVA results for the modified regression model. The p-value for the lack-of-fit term is 0.1554, the p-value for the quadratic model remains less than 0.0001, and the model exhibits an R^2 of 0.9796, an adjusted R^2 of 0.9673, an Adeq Precision of 35.0525, and a CV of 1.33%. Compared with the pre-optimization model, these metrics demonstrate improved validity, reliability, and accuracy. The optimized regression equation is provided below:

$$\alpha = 19.56 + 0.6325A + 0.6975C + 1.5775E + 0.5975AC - 0.8034A^2 + 0.5416C^2 \quad (4.20)$$

4.5.4 *Discussion*

To determine the optimal set of the three significant parameters—with the actual resting angle as the target—the optimization module in Design-Expert 13 was used to identify the parameter combination that minimizes the error between the simulated and actual resting angles. The optimal parameters were determined as follows: fertilizer-fertilizer collision recovery coefficient = 0.44, fertilizer-fertilizer rolling friction coefficient = 0.62, and fertilizer-steel static friction coefficient = 0.43. A simulation verification was conducted in EDEM 2021.2 (Altair) using these optimal parameters, while other parameters were maintained at intermediate levels. Figure 4.13 compares the simulation and experimental results. The actual resting angle was 19.95°, whereas the simulated resting angle with the optimized parameters was 20.61%—resulting in a relative error of 3.31%. This confirms that the optimal solutions for the three significant parameters are both accurate and effective.

Table 4.10 ANOVA of modified model of Box-Behnken test

Source	Sum of squares	df	F-value	*p*-value
Model	32.2	6	79.88	<0.0001
A	3.2	1	47.64	<0.0001
C	3.89	1	57.93	<0.0001
E	19.91	1	296.33	<0.0001
AC	1.43	1	21.26	0.001
A^2	2.73	1	40.57	<0.0001
C^2	1.24	1	18.43	0.0016
Residual	0.6718	10		
Lack of fit items	0.5488	6	2.97	0.1554
Pure error	0.123	4		
Cor total	32.87	16		

(a) Simulation test (b) Real tests

Fig. 4.13 Comparison between simulation test and real test. (**a**) Simulation test. (**b**) Real tests

4.6 Conclusion

1. Huachang 15-15-15 compound fertilizer was selected as the research material, with the angle of repose used as the response variable. The contact parameters of the fertilizer particles were calibrated through a combination of physical experiments and simulation tests. Plackett-Burman design, steepest ascent test, and Box-Behnken design were implemented in sequence to identify the simulated angle of repose that most closely matches the results of physical tests. The Huachang 15-15-15 compound fertilizer has a Poisson's ratio of 0.24, an elastic modulus of 2.85×10^7 Pa, a shear modulus of 1.15×10^7 Pa, and a moisture content of 5.6%—all of which serve as the basis for the calibration of simulation parameters.
2. A flexible straw model was developed by combining the Hertz-Mindlin (no-slip) contact model with the bonded-particle model (BPM). This model captures the mechanical behavior of straw and can simulate failure phenomena such as breaking, fracturing, bending, and shearing. The simulated tensile, compressive, bending, and shear responses of the straw model closely replicate the behavior of real straw. Parameter calibration and verification show that the BPM uses two types of bonds with distinct bonding parameters to connect straw particles. When the bonding parameters are set to Bond1 (2.72×10^{12} N/m^3, 1.8×10^{12} N/m^3, 4.08×10^7 Pa, 2.54×10^7 Pa) and Bond2 (5.3×10^{10} N/m^3, 3.77×10^{10} N/m^3, 7.32×10^6 Pa, 3.79×10^6 Pa), the mechanical properties of the straw model most closely resemble those of actual straw. Under these conditions, the tensile strength, compressive strength, elastic modulus, and shear strength of the constructed straw model are 29.72, 69.17, 529.69, and 6.18 MPa, respectively—all of which show more than 84% agreement with the measured values of real straw.

References

1. Zhong W, He K, Zhou Z et al (2009) Calibration of damping coefficients in discrete element simulation of particles. J Phys 58:5155–5161
2. Hao J, Long S, Li H et al (2019) Discrete element model construction and parameter calibration of cultivated yam. Trans Chin Soc Agric Eng 35:34–42

3. Wu M, Cong J, Yan Q et al (2020) Calibration and experiment of discrete element simulation parameters of peanut seed particles. Trans Chin Soc Agric Eng 36:30–38
4. Liu W, He J, Li H (2018) Simulation parameter calibration of miniature potato based on discrete element. Trans Chin Soc Agric Mach 49:125–135, 142
5. Li Y, Li F, Xue M et al (2019) Discrete element parameter calibration of wheat flour based on particle scaling. J Agric Eng 35:320–327
6. Shi L, Zhao W, Sun W (2017) Discrete element-based particle contact model and parameter calibration for agricultural soils in dry areas of Northwest China. J Agric Eng 33:181–187
7. Liu C, Wei D, Song J et al (2018) Systematization of boundary parameters for discrete element simulation of granular fertilizer. J Agric Mach 49:82–89
8. Qiu C, Zhang Q, Yan S et al (2017) Clay bioadhesion experimental study. Rock Soil Mech 42:1267–1272
9. Ren L, Chen D, Chen B (1990) Overview of soil adhesion research. Trans Chin Soc Agric Eng 1:1–7
10. Cool S, Pieters J, Mertens KC et al (2014) A simulation of the influence of spinning on the ballistic flight of spherical fertiliser grains. Comput Electron Agric 105(5):121–131
11. Xing X, Ma X, Chen L et al (2020) Physical characteristics of solid particle fertilizer experiment study. J Agric Mach Res 9:125–130
12. Chen C, Zhang S, Sun X et al (2012) Design and experiment of a self-propelled variable fertilizer spreader for upland gaps in paddy fields. J Agric Eng 28:16–21
13. Cui T, Liu J, Yang L et al (2013) Experiment and simulation of rolling friction characteristics of maize seeds based on high-speed imaging. Trans Chin Soc Agric Eng 29:34–41
14. Hu Y, Yang Y, Xiao H et al (2016) Simulation and parameter optimization of centrifugal fertilizer spreading process of tea garden fertilizer applicator. J Agric Mach 47:77–82
15. Gou Y, Li H, Wang D et al (2022) Design and simulation optimization of a small variable fertilizer spreader based on EDEM. Agric Mech Res 44(65–71):76
16. Yang L, Chen L, Zhang J et al (2018) Fertilizer sowing simulation of a variable-rate fertilizer applicator based on EDEM. IFAC-PapersOnLine 51:418–423
17. Shi L, Ma Z, Zhao W et al (2019) Discrete element simulation parameter calibration and seed row test verification of flax. Trans Chin Soc Agric Eng 35:25–33
18. Zhang R, Jiao W, Zhou J et al (2022) Parameter calibration of rice seed discrete element model with different filling particle radius. Chin J Agric Mach 51:227–235
19. Wang S, Yu Z, Aorigele et al (2022) Study on the modeling method of sunflower seed particles based on the discrete element method. Comput Electron Agric 198:107012
20. Lu C, Gao Z, Li H et al (2023) An ellipsoid modelling method for discrete element simulation of wheat seeds. Biosyst Eng 226:1–15
21. Ren J, Zhou L, Han L et al (2017) Discrete simulation of vertical spiral transportation based on particle scaling theory. Chin J Process Eng 17:936–943
22. Zou Y, Tang T, Gao Z et al (2023) Quick lime powder discrete element parameters based on the theory of particle scale calibration. China Powder Technol 29:278
23. Farid EAE, Ding WM, Ding QS et al (2015) Field investigation of a trash-board, tillage depth and low speed effect on the displacement and burial of straw. Catena 133:385–393
24. Gong YJ, Deng N, Liu DJ et al (2019) Optimization of forming process parameters and water retention performance of straw blocks. Trans Chin Soc Agric Eng 35(12):248–255
25. Chen L (2003) Theoretical and experimental studies on variable-rate fertilization in precision farming. China Agricultural University, Beijing
26. Kovács Á, Rádics JP, Kerényi G (2017) A discrete element model for agricultural decision support. In: Proceedings 31st European conference on modelling and simulation ©ECMS
27. Kweon G, Grift TE, Miclet D et al (2009) Analysis and control of uniformity by the feed gate adaptation of a granular spreader. J Biosyst Eng 34:95–105
28. Cundall PA, Strack ODL (1979) A discrete numerical mode for granular assemblies. Geotechnique 29:47–65

Chapter 5
Performance Analysis and Optimization of Centrifugal Variable-Rate Fertilizer Spreaders

J. Zheng, Man Chen, Xuekai Huang, Z. Zhao, Xiaochan Wang, and X. Yang

Abstract The accurate and uniform distribution of granular fertilizers is critical to optimizing crop yield, minimizing nutrient loss, and promoting sustainable agricultural practices. In precision agriculture, variable-rate fertilization (VRF) systems have emerged as an essential technology that adjusts fertilizer input according to real-time spatial and temporal field variability. Among various VRF mechanisms, centrifugal spreaders have gained widespread adoption due to their operational simplicity, high efficiency, and compatibility with a wide range of granular materials. However, achieving consistent spreading performance under varying working conditions remains a technical challenge, particularly due to the complex dynamics involved in particle flow, machine structure, and operational parameters. Discrete Element Method (DEM)-based simulation has become an increasingly powerful tool for analyzing and optimizing the performance of fertilizer applicators. DEM offers a high-fidelity, particle-level understanding of interactions within the spreading system, thereby enabling engineers to predict material trajectories, identify uneven distribution patterns, and propose design improvements. The studies pre-

J. Zheng · X. Huang · X. Yang
College of Engineering, Nanjing Agricultural University, Nanjing, China
e-mail: 2021212006@stu.njau.edu.cn; 2022212011@stu.njau.edu.cn

M. Chen (✉)
Nanjing Research Institute for Agricultural Mechanization, Ministry of Agriculture and Rural Affairs, Nanjing, China
e-mail: chenman@caas.cn

Z. Zhao
College of Information and Electrical Engineering, China Agricultural University, Beijing, China
e-mail: zhaozhangcau@cau.edu.cn

X. Wang
College of Engineering, Nanjing Agricultural University, Nanjing, China

Jiangsu Province Engineering Lab for Modern Facility Agriculture Technology & Equipment, Nanjing, China
e-mail: wangxiaochan@njau.edu.cn

Y. Shi et al. (eds.), *Technological Progress on Variable Rate Fertilization*, Smart Agriculture 17, https://doi.org/10.1007/978-981-95-4330-4_5

sented in this chapter focus on the numerical modeling and optimization of centrifugal variable-rate fertilizer applicators using DEM and CFD-DEM coupling methods. These investigations focus on critical engineering challenges, including the effects of disc geometry, particle characteristics, and rotational speed on distribution uniformity, with the goal of developing effective strategies for structural and operational optimization. These studies collectively address crucial engineering problems, such as the influence of disc geometry, particle properties, and rotational speed on distribution uniformity, and present effective strategies for structural and operational optimization. By integrating simulation and experimental approaches, they contribute to the theoretical basis and practical advancement of VRF technologies. This chapter aims to consolidate and critically examine their methodologies, findings, and implications, laying the groundwork for the intelligent design and optimization of next-generation fertilizer application systems.

Keywords Variable-rate technology · Centrifugal fertilizer spreader · Spreading performance · Field validation · Centrifugal variable-rate fertilizer spreader · Discrete element simulation

5.1 Introduction

As a crucial element in the implementation of precision agriculture, variable-rate fertilization is currently undergoing rapid development because of its ability to improve fertilizer use efficiency, lower production costs, and increase farmers' economic returns. Mechanized variable-rate fertilization technologies have been applied to achieve high-yield cultivation, enhanced product quality, and improved environmental protection, thereby addressing many pressing challenges.

In recent years, considerable progress has been made both domestically and internationally with grooved-wheel variable-rate fertilizer applicators. Nevertheless, centrifugal disc spreaders, which are widely recognized for their high efficiency and uniformity in application, have become increasingly favored in regions characterized by large-scale farmland. Consequently, a growing number of scholars have turned their attention to centrifugal variable-rate fertilizer spreaders. For example, Dong et al. designed a novel spreader based on the fertilizer particle motion model established by Lv et al. [1], incorporating a conical disc and spreader structure optimized through bench testing to meet the requirements of potato fertilization. Wang et al. investigated variable-rate spreaders equipped with chain conveyors [2], analyzing the structural characteristics of key components and the control system, although this design was associated with relatively complex fertilizer falling behavior and extended operational time. Hu et al. introduced a narrow spreader equipped with an offset centrifugal plate suitable for confined areas such as tea plantations, where particle dynamics and fertilizer distribution were examined using EDEM software [3], and the optimal operating parameters were determined for tea garden conditions. In general, most existing studies have focused primarily on the structural design and control optimization of centrifugal spreaders, whereas systematic research into spreading performance and the spatiotemporal behavior of fertilizer particles remains limited due to the substantial testing requirements necessary to capture such complex dynamics.

Theoretical and experimental studies on spinner-disc spreaders date back to the early 1980s [4], when the kinetic properties of granular fertilizers were analyzed through dynamic simulation models. In 2011, Coetzee et al. proposed a trajectory and dynamic model describing the motion of fertilizer particles on the disc and in flight [5]. Since then, the Discrete Element Method (DEM) has been employed to establish simulation models for spinner-disc spreaders, allowing researchers to assess how the physical properties of granular fertilizers, including particle size, friction coefficient, and restitution coefficient, affect distribution patterns [6–8]. Verification tests further confirmed the reliability and accuracy of the DEM-based models [9]. In another contribution, Grift et al. developed a feed gate control strategy utilizing an optical feedback sensor to reduce variability in uniformity [10, 11]. Simulation results demonstrated that this approach was capable of producing consistent distribution patterns at different application rates, offering superior performance compared with centrifugal spreaders equipped with alternative control systems. Additional research also examined the influence of applied flow rate and machine structure on the coefficient of variation of distribution as well as on the effective swath width [12, 13].

More recently, several systematic investigations into centrifugal variable-rate fertilizer spreaders with spinner discs have been reported. Chen et al. developed a high-clearance spreader with dual discs and identified suitable structural parameters along with a transmission device and control system [14]. Zhang et al. proposed a chain conveyor-based spreader, deriving forced motion equations for granular fertilizers and determining an appropriate hydraulic control system and mechanical configuration [2]. Lv et al. conducted kinematic analyses of fertilizer particles on a conical disc and in flight, and validated the spreading uniformity through laboratory experiments [1].

The present study seeks to address the lack of research on the spreading performance of centrifugal variable-rate fertilizer applicators, particularly under the constraints of large test areas, extended testing periods, and the need for multiple complex experiments. Building on the variable-rate fertilizer applicator for winter wheat developed by our research group using crop canopy spectral reflectance technology [14–17], a simulation model of fertilizer coverage distribution for a conical centrifugal spreader was established with the discrete element analysis software EDEM. Numerical simulations of fertilizer particle dynamics during the spreading process were conducted, and a relationship model between the coefficient of variation of transverse particle distribution and the applicator's structural parameters was developed. Using response surface analysis, an optimized design was derived and validated through testing [18, 19], significantly improving both research efficiency and the spreading uniformity of the proposed conical centrifugal variable-rate spreader. With reference to our previously developed applicator based on winter wheat canopy reflectance, this work further investigated practical issues associated with large-area fertilization, low utilization efficiency, and uneven particle distribution. The objective was to clarify the relationships between distribution variability and the applicator's working parameters. A DEM simulation model of the centrifugal spreader was constructed, and subsequent field experiments were performed to validate the simulation results. The findings confirmed the model's reliability and demonstrated improvements in spreading performance of the spreader.

5.2 Structure of the Fertilizer Spreader

5.2.1 Spreader Structure

The integral structure of the developed real-time centrifugal variable-rate fertilizer applicator, designed using crop canopy spectrum technology, is illustrated in Fig. 5.1. The system consists of a spectral sensor, vehicle-mounted control terminal (CPU), walking unit, spreading device, control system, and several auxiliary components [20, 21].

5.2.2 Operating Principle

The previously developed centrifugal variable-rate fertilizer applicator demonstrated high automation, consistent fertilizer distribution, and improved nutrient utilization, making it suitable for large-scale modern agriculture. During operation, normalized difference vegetation index (NDVI) values acquired via a GreenSeeker™

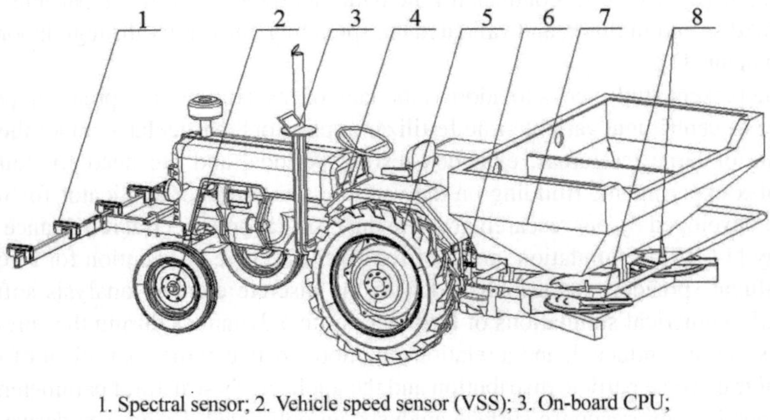

1. Spectral sensor; 2. Vehicle speed sensor (VSS); 3. On-board CPU;
4. Control system; 5. Walking system;
6. Transmission system; 7. Fertilizer can; 8. Spreader discs

Fig. 5.1 Structural diagram of the subject centrifugal variable-rate fertilizer applicator. 1. Spectral sensor; 2. Vehicle speed sensor (VSS); 3. On-board CPU; 4. Control system; 5. Walking system; 6. Transmission system; 7. Fertilizer can; 8. Spreader discs spectral detection system are transmitted in real time to the onboard CPU through a wireless serial interface module. An optimized Raun model-based decision algorithm dynamically calculates the target fertilizer rate. Feedback parameters—including walking speed, disc rotation speed, and feed gate opening (particle flow rate)—are incorporated into the control loop. The STM32 microcontroller drives a stepper motor to adjust the feed gate opening, enabling real-time, precise variable-rate fertilization.

5.2.3 Design Parameters of the Spreader Disc

The previously developed applicator measured 2300 × 1450 × 1450 mm, achieving an effective spreading swath width greater than 25 m, and was mounted on a tractor using a three-point suspension system.

The centrifugal spreader designed in this study (Fig. 5.2a) consisted mainly of a conical disc, spreading vanes, and a centrifugal distribution cover. The disc diameter was 620 mm, with a cone wall angle (α) adjustable from 0° to 20° to extend the effective swath. Two spreading vanes were positioned 180° apart, and their pitch angle (β), adjustable between 0° and 40° relative to the central generatrix, could be fine-tuned via an adjustment screw. The cylindrical distribution cover, located between the vanes, had a diameter of 240 mm, height of 80 mm, and wall thickness of 5 mm. Its base was open, with a 45 mm discharge gap between the vanes above the disc. Fertilizer particles, propelled outward by centrifugal force, exited through the cover opening and were guided by the vanes, ensuring uniform distribution along the disc edges. The fertilizer discharge regulation device (Fig. 5.2b) employed

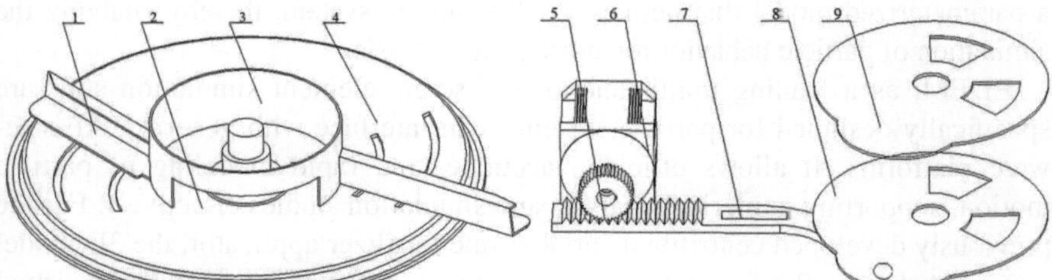

a. Centrifugal spreader disc b. Regulating device

1. Vane; 2. Output of the distribution cover; 3. Centrifugal distribution cover; 4. Conical spreader disc; 5. Pinion-rack mechanism;6. Stepper motor; 7. Link mechanism; 8. Moving plate; 9. Fixed plate

Fig. 5.2 Structural diagram of subject centrifugal spreader disc and fertilizer flow regulating device. (**a**) Centrifugal spreader disc. (**b**) Regulating device. 1. Vane; 2. Output of the distribution cover; 3. Centrifugal distribution cover; 4. Conical spreader disc; 5. Pinion-rack mechanism; 6. Stepper motor; 7. Link mechanism; 8. Moving plate; 9. Fixed plate

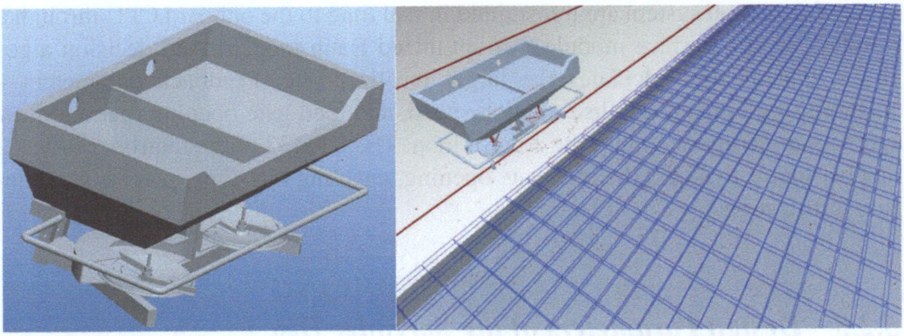

a. 3D model b. Simulation model

Fig. 5.3 D model assembly and simulation space of the subject centrifugal spreader. (**a**) 3D model. (**b**) Simulation model

a rack-and-gear mechanism driven by a stepper motor, which adjusted the overlap between holes on the moving and fixed plates to control the feed gate opening. This enabled precise regulation of fertilizer flow rates from 0 to 450 g s^{-1}, thereby managing the application rate effectively.

5.3 Materials and Methods

5.3.1 Discrete Element Model of Fertilizer Applicator

The discrete element method (DEM) represents an advanced numerical technique for analyzing the dynamics of complex discrete systems. It operates by developing a parameterized model that defines a solid particle system, thereby enabling the simulation of particle behavior for subsequent analysis.

EDEM, as a leading multi-functional discrete element simulation software specifically designed for particle systems, can interface with general CAE software platforms. It allows efficient, accurate, and rapid modeling of particle motion, supporting numerical analysis and simulation studies effectively. For the previously developed centrifugal variable-rate fertilizer applicator, the 3D model was created using Pro/E, a parametric modeling tool. After incorporating a virtual driver and applying assembly constraints, the model was exported to the EDEM environment in ".igs" format. The assembled model components are shown in Fig. 5.3.

Table 5.1 Material properties of fertilizer particles

Parameter	Particle	Ground	Steel
Poisson's ratio	0.24	0.51	0.3
Shear modulus (MPa)	10.76	1.1×10^2	7×10^4
Density (kg m^{-3})	1.38×10^3	1.25×10^3	7.8×10^3
Contact mechanical parameters	Elastic recovery coefficient	Particle to particle	0.27
		Ground to particle	0.06
		Steel to particle	0.29
	Friction coefficient	Particle to particle	0.26
		Ground to particle	1.20
		Steel to particle	0.59
	Static friction coefficient	Particle to particle	0.34
		Ground to particle	1.13
		Steel to particle	0.48
Average diameter of particle (mm)		4.02	

5.3.2 Selection of the Fertilizer Characteristic Parameters

A typical KCl-based compound fertilizer was selected as the experimental material. For simulation purposes, the fertilizer particles were considered spherical, randomly sampled, and their physical properties characterized. The contact parameters describing interactions among fertilizer particles, the spreader disc/vanes (steel), and the ground are summarized in Table 5.1. The EDEM default Hertz–Mindlin contact model without sliding was adopted [22]. A particle factory was configured to generate 20 kg of statically distributed particles. The centrifugal disc was set to rotate at 600 r min^{-1}, the fertilizer outlet flow rate at 300 g s^{-1}, and the simulation duration at 10 s. These settings were consistent with actual field equipment to ensure accurate simulation outcomes. These parameters corresponded to actual field equipment settings, ensuring the accuracy of the simulation results

5.3.3 Design of Simulation Experiment

The discrete element model of the fertilizer applicator equipped with a conical spreader was imported into the calculation module of EDEM (v2.2). Based on the theoretical kinematics and dynamics analysis of particles outlined in Sect. 5.2.2, several structural parameters—specifically, vane pitch angle, disc cone angle, and disc height (distance from disc to ground)—were identified as critical factors influencing particle trajectories and, consequently, the uniformity of fertilizer distribution. Following the static test protocol for centrifugal spreaders in ISO 5690-2 [13, 18] and ASAE S341.2 [18, 23], a virtual test field measuring 30 × 14 m was established, containing 2100 collection pans arranged in 35 rows × 60 columns, each with dimensions of 50.8 × 40.6 × 10.2 cm. The cumulative particle mass within each

Table 5.2 Test factors and values of fertilizer spreader simulation

Test factor		Coded value					Interval Δ_i
		$(-\gamma)$	(-1)	0	$(+1)$	$(+\gamma)$	
A	Conical angle of disc α(°)	0	5°	10°	15°	20°	5°
B	Pitch angle of vane β(°)	−30°	−15°	0°	15°	30°	15°
C	Height of disc bench h(cm)	50	65	80	95	110	15

column of 35 pans was summed to form a 1 × 60 fertilizer collection matrix. This matrix was then analyzed to represent the single-pass transverse distribution of fertilizer as the applicator traversed the centerline of the test area at a designated ground speed. The coefficient of variation (CV) of the transverse particle distribution was computed to quantify the spreading performance of the centrifugal applicator [24, 25]. The equations for the coefficient of variation are as follows:

$$\begin{cases} CV = \dfrac{SD}{\overline{X}} \times 100\% \\ SD = \left\{ \sum_{i=1}^{n} \left[\left(X_i - \overline{X} \right)^2 \right] / (n-1) \right\}^{\frac{1}{2}} \\ \overline{X} = \dfrac{1}{n} \sum_{i=1}^{n} X_i \qquad (i = 1,2,3,\ldots,n) \end{cases} \tag{5.1}$$

where CV is the coefficient of variation of particle distribution, SD is the standard deviation of fertilizer amount, $\overline{X}$ is the mean amount of particles, Xi is the amount of particles in the ith collection pan (g), and n is the number of collection pans.

To assess the distribution of fertilizer particle mass across the test field, focusing on vane pitch angle, disc cone angle, and disc bench height as key factors, the coefficient of variation (CV) of the transverse particle distribution was used as the evaluation metric. Consequently, a quadratic regression orthogonal experiment was designed, combining the three factors at five levels each, to investigate the distribution uniformity of the developed conical centrifugal variable-rate spreader. Referring to the performance parameters of comparable variable-rate fertilizer applicators, the values and ranges for each factor are presented in Table 5.2, and all 23 experimental schemes are listed in Table 5.3.

Table 5.3 Fertilizer spreader simulation testing scheme and results

Test	Factors			Coefficient of variation C_V (%)
	A	B	C	
1	1	1	1	11.17
2	1	1	−1	11.84
3	1	−1	1	15.39
4	1	−1	−1	15.92
5	−1	1	1	11.46
6	−1	1	−1	11.73
7	−1	−1	1	15.54
8	−1	−1	−1	15.91
9	−1.682	0	0	13.22
10	1.682	0	0	13.45
11	0	−1.682	0	16.87
12	0	1.682	0	14.24
13	0	0	−1.682	14.86
14	0	0	1.682	12.53
15	0	0	0	12.25
16	0	0	0	11.67
17	0	0	0	12.39
18	0	0	0	12.06
19	0	0	0	12.44
20	0	0	0	12.58
21	0	0	0	11.94
22	0	0	0	12.27
23	0	0	0	12.68

5.3.4 Field Test Method

To verify the accuracy of the simulation model and the rationality of the virtual structural optimization, field tests were conducted at the test station of Yancheng Yanhai Tractor Manufacturing Co., LTD, located in Jiangsu Province, in April 2017. During the tests, the weather was clear, with the temperature ranging from 15 to 25 °C and the wind speed kept below 2.0 m/s—this wind speed met the test wind speed requirements specified in the ASAE standard. Additionally, the ground surface was relatively flat, the test area covered 450 m^2, the air relative humidity was 42%, and the soil absolute moisture content was 21%. The test methods and evaluation indices strictly adhered to the requirements of standards GB/T 5262-2008 (General Rules for the Determination of Agricultural Mechanical Test Conditions), ISO 5690, and ASAE S341.2. Several images captured during the field experiment of the spreader are presented in Fig. 5.4.

To minimize fertilizer wastage, a plastic film was laid over the test area (measuring 30 m × 14 m). Meanwhile, soft cotton cloths were placed in each collection pan to mitigate errors induced by the bouncing effect of fertilizer granules. A total of

Fig. 5.4 Test of spreading fertilizer in field

150 collection pans—each with dimensions of 50.6 cm × 40.8 cm × 10.2 cm—were arranged in a 15-row × 10-column matrix, with intervals of 1.5 m along the column direction and 0.8 m along the row direction. To characterize the transverse distribution of fertilizer, the mass of fertilizer granules in each collection pan (which were labeled and subsequently stored in bags) was measured. Subsequently, the fertilizer quantity was adjusted to fill half of the can, and the spreader was calibrated in accordance with the specific conditions of each test. After achieving stable and smooth operation in the stable region, the spreader was operated over the measurement area (i.e., the area where the collection pans were placed). At the end of each row, the spreader was stopped to guarantee the accuracy of the test in defining the target region. During the trial, parameters including the vane pitch angle, disc rotational speed, distributing flow rate, and spreading height were adjusted. A model describing the fertilizer particle distribution was then established with consideration of the effective swath width, and this model was utilized to calculate the coefficient of variation for each test case. The experimental results obtained were compared with the simulation results. The walking trajectory of the spreader and the array of collection pans are presented in Fig. 5.5.

5.4 Results and Analysis

5.4.1 Results of Simulation and Variance Analysis

The simulation test program described in Sect. 5.3.3 was carried out, and the corresponding results are presented in Table 5.3. A quadratic regression analysis of these outcomes was conducted using the Design Expert 8.0.6 software, establishing a regression model to characterize the relationship between influence factors and evaluation indices.

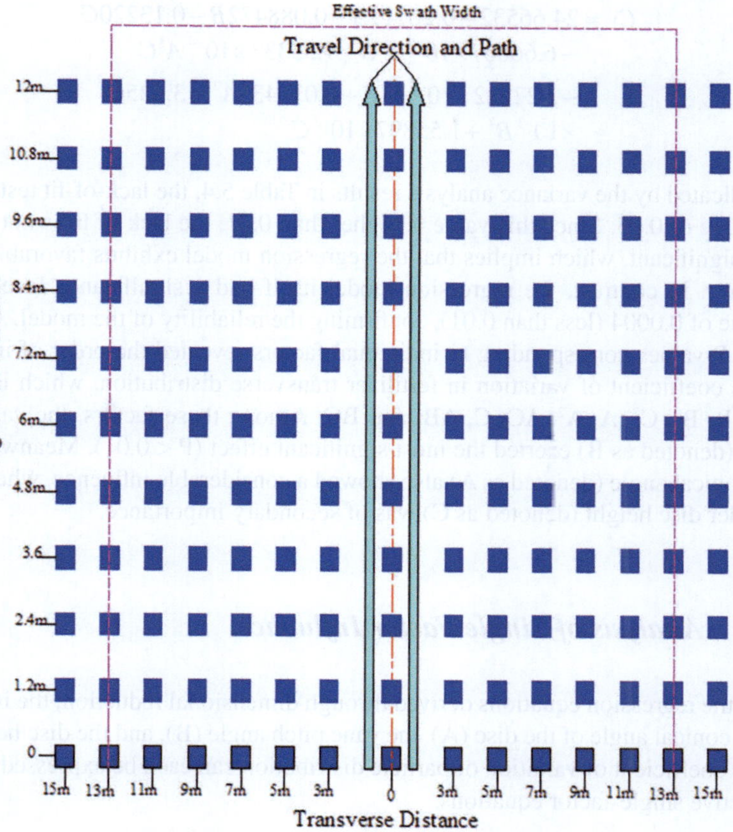

Fig. 5.5 Schematic diagram of collection pan matrix

Table 5.4 Analysis of variance of fertilizer spreader test factors on the outcome of the simulation

Source of variation	Sum of squares	DOF	Mean square	F values	P values
Model	53.74	9	5.94	8.24	0.0004
A	2.64	1	2.64	3.66	0.0780
B	29.76	1	29.76	41.26	<0.0001
C	0.00012	1	0.00012	0.00017	0.9687
AB	0.00002	1	0.00002	0.00002	0.9870
AC	0.039	1	0.039	0.054	0.8193
BC	0.00002	1	0.00002	0.00002	0.9870
A^2	1.91	1	1.91	2.65	0.1275
B^2	18.71	1	18.71	25.93	0.0002
C^2	3.46	1	3.46	4.79	0.0474
Residual	9.38	13	0.72		
Lack of fit	8.56	5	1.71	16.73	0.05
Error	0.82	8	0.10		
Total	62.85	22			

$$Cv = 24.66532 - 0.26719A - 0.088472B - 0.13220C \\ - 6.66667 \times 10^{-5} A^{*}B - 9.33333 \times 10^{-4} A^{*}C \\ - 2.22222 \times 10^{-5} B^{*}C + 0.010431A^{2} + 3.62564 \\ \times 10^{-3} B^{2} + 1.55897 \times 10^{-3} C^{2} \tag{5.2}$$

As indicated by the variance analysis results in Table 5.4, the lack-of-fit test yielded a P-value of 0.05. Since this value is higher than 0.01, the lack of fit is not statistically significant, which implies that the regression model exhibits favorable fitting accuracy. In contrast, the regression model itself had a significance level with a P-value of 0.0004 (less than 0.01), confirming the reliability of the model. Analysis of the P-values corresponding to individual factors revealed the order of influence on the coefficient of variation in fertilizer transverse distribution, which is as follows: B, B^2, C^2, A, A^2, AC, C, AB, and BC. Among these factors, the vane pitch angle (denoted as B) exerted the most significant effect ($P < 0.01$). Meanwhile, the disc conical angle (denoted as A) also showed a considerable influence, whereas the spreader disc height (denoted as C) was of secondary importance.

5.4.2 Analysis of Single Factor Influence

From the regression equations derived through dimensional reduction, the influence of the conical angle of the disc (A), the vane pitch angle (B), and the disc height (C) on the coefficient of variation of particle distribution can each be expressed by their respective single-factor equations.

$$\begin{aligned} Cv_A &= 24.66532 - 0.26719A + 0.010431A^2 \\ Cv_B &= 24.66532 - 0.088472B + 3.62564 \times 10^{-3} B^2 \\ Cv_C &= 24.66532 - 0.13220C + 1.55897 \times 10^{-3} C^2 \end{aligned} \tag{5.3}$$

Based on the single-factor regression equations, the impacts of each parameter on the coefficient of variation of fertilizer distribution are presented in Fig. 5.6. As the disc conical angle increases, the coefficient of variation first rises slightly and subsequently decreases—this trend indicates that a specific conical angle is capable of improving the effective spreading width. However, once this angle surpasses a threshold value, the coefficient of variation of particle distribution increases gradually. The minimum coefficient of variation is achieved when the disc conical angle α is set to 15°. As the vane pitch angle increases, the coefficient of variation first rises significantly and then decreases. This phenomenon implies that a moderate positive pitch angle enhances spreading uniformity; in contrast, an excessively large pitch angle leads to the accumulation of fertilizer particles in localized areas, thereby compromising distribution uniformity. The minimum coefficient of variation is observed when the vane pitch angle β is 15°. Furthermore, the coefficient of variation increases consistently as the spreading height increases. This trend indicates

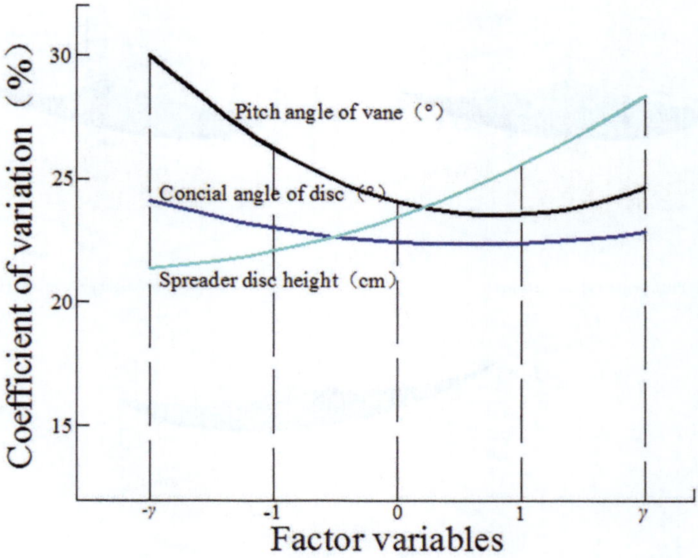

Fig. 5.6 Effect of the change in single factors on the coefficient of variation of fertilizer distribution

that although a higher disc position expands the effective spreading width, it also results in an increased variation in particle distribution.

5.4.3 *Multi-factor Effect Analysis and Optimization Using the Response Surface Method*

A clearer understanding of the interactive effects of these structural parameters on the fertilizer particle distribution can be obtained through multifactor dimensionality reduction and the response surface method (RSM). In this approach, one factor is held constant while the other two vary, generating three-dimensional response plots (Fig. 5.7).

As depicted in Fig. 5.7a, under a fixed spreader disc height, increasing both the disc conical angle and the vane pitch angle causes the coefficient of variation of fertilizer particles' transverse distribution to first decrease and then increase gradually. The slope of the response surface suggests that the vane pitch angle (β) has a more significant impact on the coefficient of variation than the disc conical angle (α). Relatively low values of the coefficient of variation are achieved when α ranges from 10° to 15° and β ranges from 10° to 20°. In Fig. 5.7b, with the disc conical angle kept constant, a gradual increase in both the vane pitch angle and the spreader disc height leads to a slight initial decrease in the coefficient of variation, followed by an upward trend. The response surface reveals that the rate of change in the direction of β is greater than that along the spreader disc height (H), indicating that the

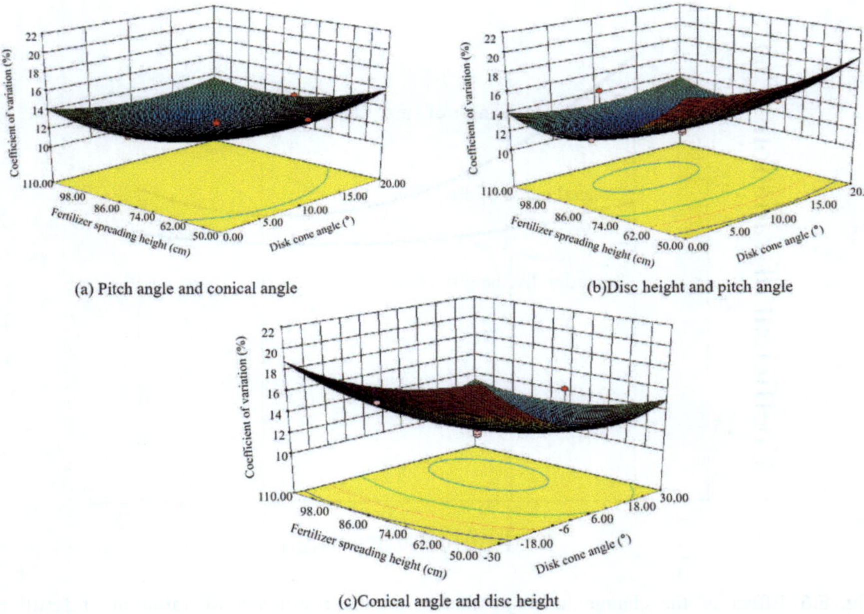

Fig. 5.7 Surface plots illustrating the effect of factor interaction on the coefficient of variation. (**a**) Pitch angle and conical angle. (**b**) Disc height and pitch angle. (**c**) Conical angle and disc height

vane pitch angle acts as the dominant factor in this interaction. The minimum coefficient of variation is observed when H is between 80 and 100 cm and β is between 10° and 15°. In Fig. 5.7c, with the vane pitch angle held constant, the coefficient of variation of fertilizer particle distribution also shows a trend of initial decrease followed by increase as the spreader disc height and disc conical angle increase. In this interaction, α exerts a slightly more prominent effect on the coefficient of variation than H, and the optimal values are determined to be α = 10–15° and H = 80–100 cm.

To identify the optimal structural parameters of the developed conical centrifugal fertilizer spreader, response surface methodology (RSM) was employed to optimize the parameters for minimizing the coefficient of variation, based on the above two-factor interaction analyses. By considering the three response surfaces presented in Fig. 5.7, the optimal parameter combination predicted by RSM is α = 10.35°, β = 12.58°, and H = 88.88 cm. This combination results in a minimum coefficient of variation of 11.55% for the transverse distribution of fertilizer particles within the effective spreader swath width.

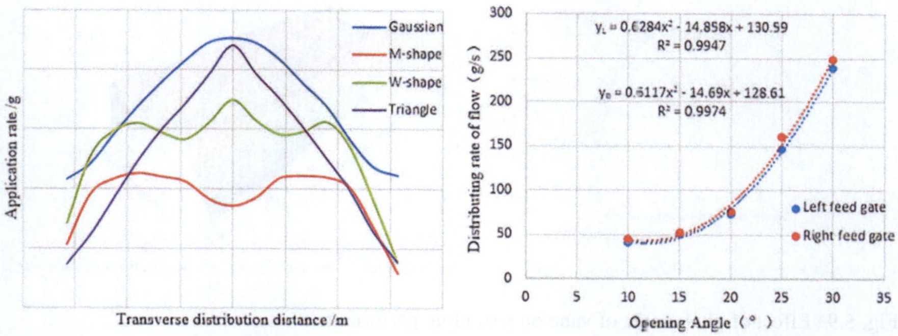

a. Distribution shape b. Calibration of flow rate

Fig. 5.8 Transverse distribution shape of spreading particle and calibration of particle flow rate. (**a**) Distribution shape. (**b**) Calibration of flow rate

5.4.4 Transverse Distribution Shape and Calibration of Flow Rate

Based on the operational characteristics of the spreader, the transverse distribution of applied fertilizer particles is symmetric with respect to the axis of the fertilizer applicator. Moreover, due to variations in particle physical properties, spreader operational parameters, and environmental conditions, the distribution patterns mainly exhibit Gaussian, triangular, M-type, and W-type shapes, as shown in Fig. 5.8a. Among these patterns, the Gaussian distribution is optimal, as it minimizes the coefficient of variation (Cv) within the effective swath width (with steep edge distributions excluded). In contrast, the Cv values for W-type and M-type distributions within the effective swath width are relatively higher.

The feed flow rates of the two fertilizer cans need to be calibrated to accurately determine the distinct fertilizer application characteristics. A commonly used compound fertilizer was selected for the test, and feed flow rates were assessed under different feed opening angles (for calibration) of 10°, 15°, 20°, 25°, and 30°. During the fertilizer trials, collection bags were placed beneath each feed gate, and each test was conducted for 10 s. The distributed flow rate was calculated indirectly by measuring the mass of fertilizer in a single test bag, with each test repeated three times. The calibration results are presented in Fig. 5.8b, which show that the test deviation did not exceed 12.14%. The results also reveal a quadratic polynomial positive correlation between the flow rate and the feed gate opening angle, reflecting the tendency of fertilizer particles to distribute easily.

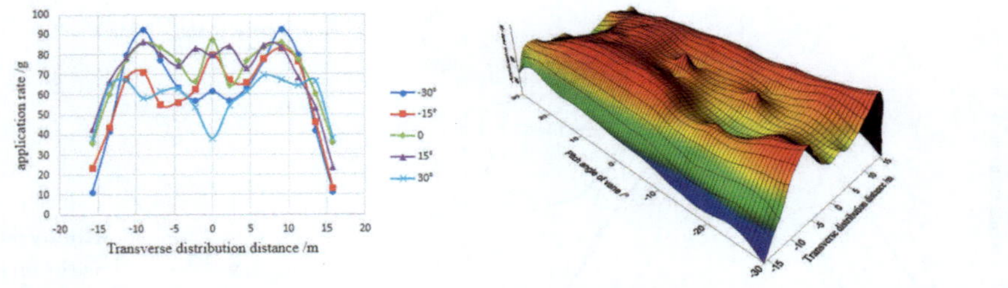

(a) Transverse distribution patterns under different vane pitch angles. (b) 3D response surface of spreading uniformity versus pitch angle and transverse distance

Fig. 5.9 Effect of pitch angle of vane on spreading performance

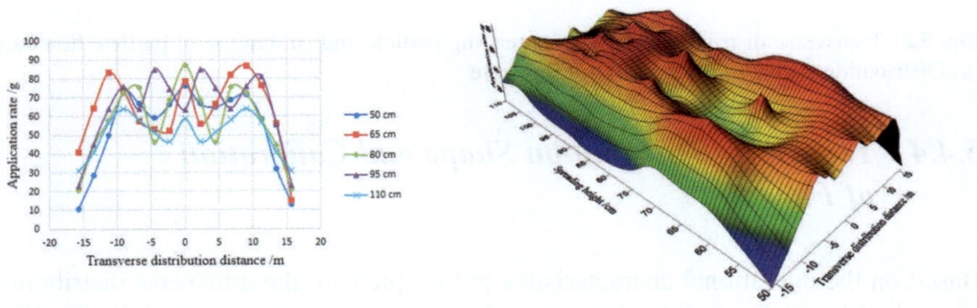

(a) Variation in transverse distribution uniformity across disc heights. (b) Identification of the optimal disc height for achieving maximum uniformity.

Fig. 5.10 Effect of disc height on spreading performance

5.4.5 Test Results and Analysis

5.4.5.1 Effects of Pitch Angle of Vane

To evaluate the influence of different vane pitch angles on spreader performance, tests were conducted under a fixed fertilizer application rate of 225 kg/hm^2. The test results are shown in Fig. 5.9a. As the vane pitch angle increased, the transverse distribution pattern of fertilizer particles changed irregularly. When $\beta = -30°$, the distribution was predominantly M-type (with a lower fertilizer rate at the center), featuring a standard deviation of 16.54 and a maximum Cv of 24.33% within the effective swath width. In contrast, when $\beta = 15°$, the particle distribution approached a W-type (with a higher fertilizer rate at the center), with a standard deviation of 9.36 and a minimum Cv of 12.23%. As illustrated in Fig. 5.9b, the Cv decreased gradually as β increased from −30° to 15°, whereas a further increase in β from 15° to 30° led to a gradual rise in Cv. Thus, $\beta = 15°$ corresponds to the minimum Cv and better spreading consistency.

Note: Here, the long axis represents the vane pitch angle, the broad axis represents the transverse distribution distance, and the high axis represents the application rate.

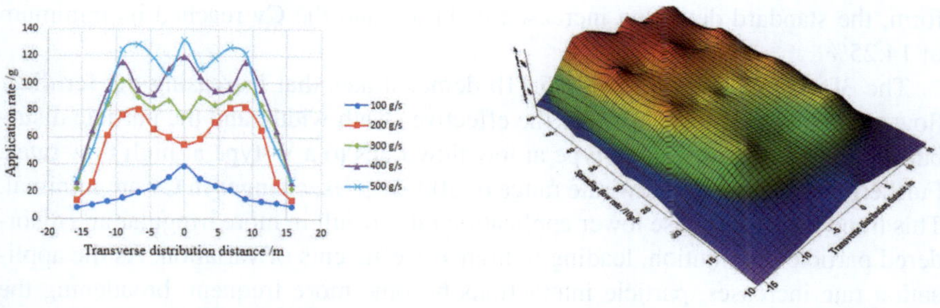

(a) Transverse distribution patterns under different single-orifice flow rates. (b) Three-dimensional visualization of fertilizer spatial distribution.

Fig. 5.11 Effect of flow rate on spreading performance

5.4.5.2 Effects of Disc Height

The influence of varying spreader disc heights on spreader performance was also investigated, with the test results presented in Fig. 5.10a. The particle distribution shape changed irregularly as the disc height increased, while the coefficient of particle distribution decreased with increasing disc height. At h = 110 cm, the standard deviation of the fertilizer application rate was 7.91, and the Cv within the effective swath width reached its minimum value of 14.81%. Conversely, at h = 60 cm, the standard deviation was 14.98, and the Cv attained its maximum value of 24.88%. Further analysis using the 3D model in Fig. 5.10b indicates that the effective swath width of the fertilizer particle distribution expands with increasing spreader disc height. This phenomenon is likely due to the increased kinematic velocity of particles, which enhances the effective swath width. However, under such conditions, the localized application rate decreases. Therefore, to achieve both an effective site-specific fertilizer rate and a sufficient swath width, a spreader disc height of 95 cm was selected, resulting in a coefficient of variation of 16.71% and improved spreading uniformity.

Note: Here, the long axis represents the disc height, the broad axis represents the transverse distribution distance, and the high axis represents the application rate.

5.4.5.3 Effects of Flow Rate

Additionally, the influence of varying fertilizer flow rates on spreader performance was examined, with the experimental results shown in Fig. 5.11a. The distribution pattern of fertilizer particles changed significantly with different flow rates. Initially, the Cv decreased sharply and then gradually stabilized; however, the Cv remained largely unaffected by further increases in the fertilizer application rate. At q = 100 g/s, the particle distribution exhibited a triangular shape, with a standard deviation of 8.71 for the application rate and the highest Cv (42.51%) within the effective swath width. In contrast, at q = 300 g/s, the distribution took an M-type form, the standard deviation increased to 11.92, and the Cv reached its minimum of 14.25%.

The 3D representation in Fig. 5.11b demonstrates that increasing the fertilizer flow rate leads to an expansion of the effective swath width, and the particle distribution transitions from an M-type at low flow rates to a W-type at high flow rates. Furthermore, within the flow rate range of 300–500 g/s, changes in Cv are minimal. This trend occurs because lower application rates result in more irregular and disordered particle distribution, leading to higher coefficients of variation. As the application rate increases, particle interactions become more frequent, broadening the effective swath width. Once the application rate exceeds a certain threshold, some fertilizer accumulates near the center of the dual spreader discs, altering the distribution shape to a W-type. Based on a target fertilizer application rate of 225 kg/hm^2, the corresponding particle flow rate was selected as 300 g/s.

Note: Here, the long axis represents the spreading flow rate, the broad axis represents the transverse distribution distance, and the high axis represents the application rate.

5.4.5.4 Effects of Disc Rotation Speed

The influence of different disc rotation speeds on spreader performance was investigated, with the experimental results presented in Fig. 5.12a. As the disc rotation speed increased, the fertilizer particle distribution gradually transformed from a triangular shape to a W-shape, and eventually to an M-shape, while the effective swath width expanded. At ω = 300 rpm, the standard deviation of the fertilizer application rate reached 40.02, and the Cv within the effective swath width was the highest at 56.14%. In contrast, at ω = 900 rpm, the standard deviation decreased to 6.60, and the Cv reached its minimum of 12.74%. As shown in Fig. 5.12b, the Cv decreases with increasing ω, while the effective swath width of the spreader expands. This behavior can be attributed to the low centrifugal force and kinematic velocity at low rotational speeds, which reduce the effective swath width and cause granular fertilizer to accumulate near the disc—leading to uneven distribution and higher Cv. At

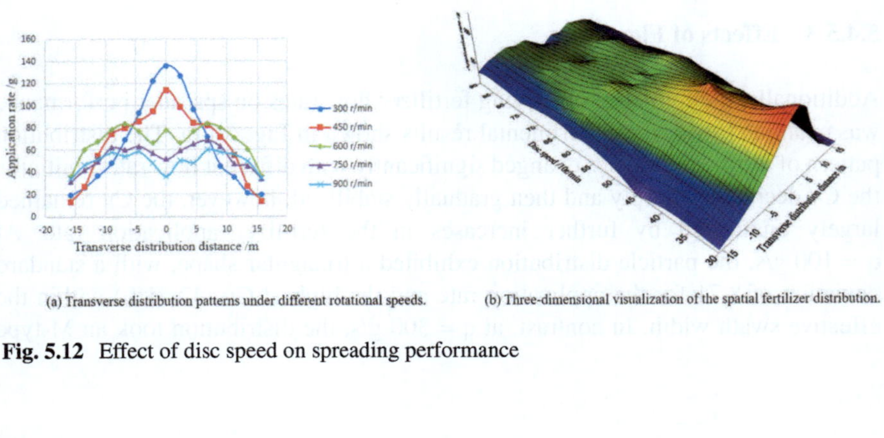

(a) Transverse distribution patterns under different rotational speeds. (b) Three-dimensional visualization of the spatial fertilizer distribution.

Fig. 5.12 Effect of disc speed on spreading performance

higher rotational speeds, centrifugal force increases, expanding the swath width and promoting more uniform particle distribution. Under a constant feed-gate flow rate, the application rate in the concentrated area is reduced. For speeds above 600 rpm, the changes in particle distribution pattern become less predictable, and the Cv remains largely constant below 15% [26, 27]. Thus, 600 r/min was considered the critical rotational speed, at which particle distribution variation is minimal and uniformity is optimized.

Note: Here, the long axis represents the disc speed, the broad axis represents the transverse distribution distance, and the high axis represents the application rate.

5.4.6 Validation Test for Optimal Parameters

Through the aforementioned analysis, a superior set of working parameters for the spreader was identified for field validation tests, including a flow rate q = 300 g/s, vane pitch angle β = 15°, disc height h = 95 cm, and disc rotational speed ω = 600 r/min. Each test was repeated three times to calculate the mean parameter values; the purpose of this design was to verify the rationality and accuracy of the simulation model, as well as to confirm the spreader's performance under the specified parameter set. Three randomly selected field tests are illustrated in Fig. 5.13.

Note: Here the long axis represents the longitudinal distance, the broad axis represents the transverse distribution distance and the high axis represents the application rate.

The longitudinal test results are summarized in Fig. 5.13 and Table 5.5. Minor irregular peaks were observed, which can be attributed to the good flowability of the

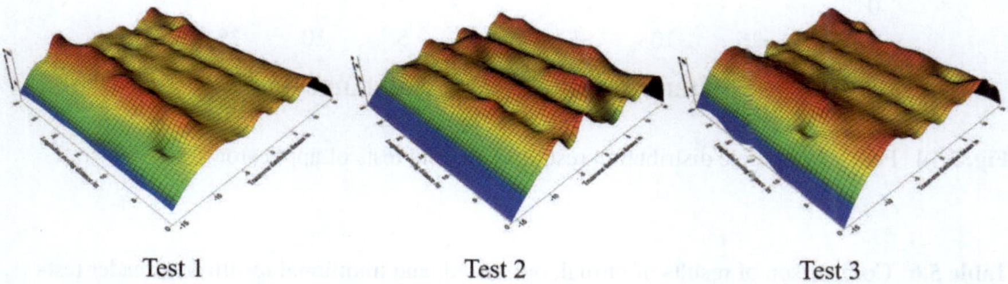

Fig. 5.13 Transverse distribution of field test by optimized working parameters

Table 5.5 Test result of field comparison test

Test	Spreader width (m)	Application rate (g/m²)	Variable coefficient(%)			Largest deviation	Relative error (%)
			1	2	3		
Simulation	22.5–26.7	66.7–84.1	12.23	16.71	14.25	4.48	10.66
Field	23.1–27.4	72.6–86.5	14.18	18.93	17.12	4.75	

fertilizer granules and their dynamic shear effects. For the optimal parameter combination, the average effective swath width, mean fertilizer application rate, average Cv within the effective swath width, and maximum test deviation were approximately 25.2 m, 79.6 g per unit area, 16.74%, and 4.75%, respectively. These values satisfy both the operational requirements of standard centrifugal fertilizer spreaders and the criteria specified in the "Technical Specification of Quality Evaluation for Fertilizing Machinery" (NY/T 1003-2006), which mandates a Cv ≤ 20%. When compared with the simulation results, the field test results exhibited a relative mean error of 10.66%, confirming the simulation model's accuracy and validity. This finding demonstrates that the simulation approach can effectively replicate the field performance of the centrifugal variable-rate fertilizer applicator, thereby reducing labor demands.

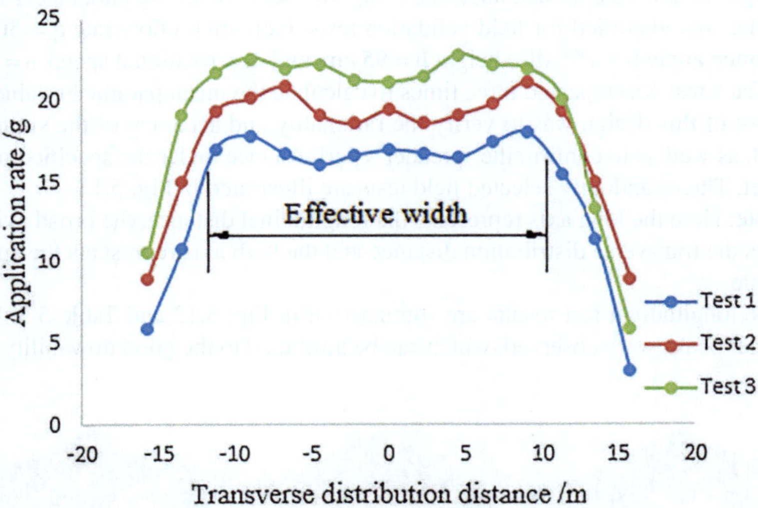

Fig. 5.14 Fertilizer particle distribution results from field tests of applicator

Table 5.6 Comparison of results of virtual, optimized, and traditional fertilizer spreader tests

	Coefficient of variation (%)										Mean error (%)
Test	1	2	3	4	5	6	7	8	9	10	
Virtual test	11.48	11.32	11.16	11.83	11.69	12.07	11.58	11.46	11.29	11.73	11.78
Optimization test	12.16	13.53	11.69	13.27	14.62	13.51	12.49	12.75	14.68	12.35	
Traditional test	15.67	16.82	14.39	17.23	18.14	16.39	15.62	17.48	16.75	15.94	−20.3

5.5 Field Verification of Fertilizer Applicator

5.5.1 *Analysis of Field Test Results*

Figure 5.14 presents the field validation results of the previously developed variable-rate fertilizer applicator under the optimal parameter set derived from response surface regression analysis, including three representative field experiments. Table 5.6 provides a comparison of these results with data from discrete element simulations and tests conducted using a conventional spreader (without a centrifugal distribution cover). Both sets of field tests were performed using identical parameters, and the layout of the 150 collection boxes adopted in the tests was consistent with that used in the simulation setup.

A comparison of the test results reveals that the previously developed conical centrifugal variable-rate fertilizer applicator—with its structural parameters optimized via the response surface method—achieved a uniform coefficient of variation for the transverse particle distribution. The average Cv measured for this centrifugal spreader was 13.1%, which represents a 20.3% reduction in variation compared to the conventional spreader lacking a centrifugal distribution cover. This Cv value complies with the uniformity requirements specified in relevant technical standards for fertilizer application (Cv $\leq$ 20%) [28]. When the field performance of the applicator was compared with the simulation results, the relative error was less than 11.78%, confirming the accuracy of both the predictive model and the discrete element simulation. Thus, the applicator evaluated in this study exhibits stable spreading performance, which meets the requirements for efficient fertilizer distribution.

5.6 Conclusions

1. A conical spreader disc with a centrifugal fertilizer distribution cover was developed using the previously designed centrifugal variable-rate fertilizer applicator based on spectrum technology. A discrete element simulation model of this applicator was established in EDEM software, enabling a detailed numerical assessment of its spreading performance.
2. A factor combination testing program, designed via quadratic regression orthogonal rotation, was employed in the simulation experiments to evaluate the relationships and interaction effects among the structural parameters of the spreader on the coefficient of variation of transverse particle distribution. Response surface regression and variance analysis subsequently identified the hierarchy of factors influencing the uniformity of fertilizer distribution.
3. Validation tests confirmed that the variable-rate fertilizer applicator, with structural parameters optimized through response surface regression, achieved improved spreading uniformity. Field performance closely matched simulation results, verifying the predictive capability of the model and the accuracy of the discrete element simulation.

The results of this study demonstrate that the developed simulation model can reliably reproduce the field spreading performance of the centrifugal variable-rate fertilizer applicator, assess the uniformity and consistency of fertilizer distribution, reduce experimental workload, provide reference values for applicator design, and support the improvement of conventional empirical fertilization practices. Future work will focus on determining precise fertilizer application rates for centrifugal variable-rate applicators, accounting for the complex operational conditions of agricultural machinery and the varying nutrient requirements of crops [2, 17, 19, 25, 29].

References

1. Lv J, Shang Q, Yang Y et al (2016) Performance analysis and experiment on granular fertilizer spreader with cone disc. Trans Chin Soc Agric Eng 32(11):16–24
2. Zhang R, Wang X, Zhao CJ et al (2012) Design and experiment of variable rate fertilizer spreader with conveyor chain. Trans Chin Soc Agric Eng 28(6):20–25
3. Hu YG, Yang Y, Xiao H et al (2016) Simulation and parameter optimization of centrifugal fertilizer spreader for tea plants. Trans Chin Soc Agric Mach 47(5):77–82
4. Olieslagers R, Ramon H, Baerdemaeker JD (1996) Calculation of fertilizer distribution patterns from a spinning disc spreader by means of a simulation model. J Agric Eng Res 63(2):137–152
5. Coetzee CJ, Lombard SG (2011) Discrete element method modelling of a centrifugal fertiliser spreader. Biosyst Eng 109(4):308–325
6. Dintwa E, Van LP, Olieslagers R et al (2004) Model for simulation of particle flow on a centrifugal fertiliser spreader. Biosyst Eng 87(4):407–415
7. Liedekerke PV, Tijskens E, Dintwa E et al (2009) DEM simulations of the particle flow on a centrifugal fertilizer spreader. Powder Technol 190(3):348–360
8. Aphale A, Bolander N, Park J et al (2003) Granular fertiliser particle dynamics on and off a spinner spreader. Biosyst Eng 85(3):319–329
9. Liedekerke PV, Tijskens E, Dintwa E et al (2006) A discrete element model for simulation of a spinning disc fertilizer spreader I. Single particle simulations. Commun Agric Appl Biol Sci 170(2):71–85
10. Kweon G, Grift TE, Miclet D et al (2009) Analysis and control of uniformity by the feed gate adaptation of a granular spreader. Biosyst Eng 34(2):95–105
11. Grift TE (2000) Spread pattern analysis tool (SPAT): I. Development and theoretical examples. Trans ASAE 43(6):1341–1350
12. Fulton JP, Shearer SA, Chabra G et al (2001) Performance assessment and model development of a variable-rate, spinner-disc fertilizer applicator. Trans ASAE 44(5):1071–1081
13. ISO. Equipment for distributing fertilizers—test methods—part 2: fertilizer distributors in lines. ISO5690-2-1984. https://www.iso.org/standard/11803.html
14. Chen M, Shi Y, Wang X et al (2015) Design and experiment of variable rate fertilizer applicator based on crop canopy spectral reflectance. Trans Chin Soc Agric Mach 46(5):26–32
15. Shi Y, Chen M, Wang X et al (2015) Design and experiment of precision fertilizer applicator actuator of rice and wheat. J South China Agric Univ 36(6):119–124
16. Shi Y, Chen M, Wang X et al (2017) Analysis and experiment of fertilizing performance for precision fertilizer applicator in rice and wheat. Trans Chin Soc Agric Mach 48(7):97–103
17. Raun WR, Solie JB, Stone ML et al (2005) Optical sensor-based algorithm for crop nitrogen fertilization. Soil Sci Plant Anal 36(19–20):2759–2781
18. Chen S, Zhang S, Sun X et al (2012) Design and experiment of self-propelled high-ground-clearance spreader for paddy variable-rate fertilization. Trans Chin Soc Agric Eng 28(11):16–21

19. Wang GW, Yan L, Chen GF (2009) Comprehensive evaluation of effect of variable rate fertilization on spatial variability of soil nutrients. Trans Chin Soc Agric Eng 25(10):82–85
20. Liu Y (2012) Key technology on variable rate application system. Chinese Academy of Agricultural Mechanization Sciences, Beijing
21. Zhang R, Wang X, Zhou J et al (2012) Breadth control of scatter system for fertilizer spreader. Trans Chin Soc Agric Mach S1:39–43
22. Malone KF, Xu B (2008) Determination of contact parameters for discrete element method simulations of granular systems. Particuology 6(6):521–528
23. ASAE (2009) Procedure for measuring distribution uniformity and calibrating granular broadcast spreaders. S341.4. http://elibrary.asabe.org/azdez.asp?JID=2&AID=29220&CID=s2000&T=2
24. Jones JR, Lawrence HG, Yule IJ (2008) A statistical comparison of international fertiliser spreader test methods—confidence in bout width calculations. Powder Technol 184(3):337–351
25. Grift TE, Crespi CM (2008) Estimation of the flow rate of free falling granular particles using a Poisson model in time. Biosyst Eng 101(1):36–41
26. Cool S, Pieters J, Mertens KC et al (2014) A simulation of the influence of spinning on the ballistic flight of spherical fertiliser grains. Comput Electron Agric 105:121–131
27. Roblin P, Barrow D (2000) Microsystems technology for remote monitoring and control in sustainable agricultural practices. J Environ Monit 2(5):385–392
28. Blackmore S (1996) An information system for precision farming. In: Brighton crop protection conference on pests and diseases. British Crop Protect Council, pp 1207–1214
29. Torán Z (2015) A uniformity controlled granular fertilizer spreader. University of Illinois, Urbana-Champaign

Chapter 6
Performance Evaluation and Lag Compensation of Actuators in Variable-Rate Fertilizer Spreaders

D. Wang, Z. Zhao, Xuekai Huang, L. Xu, Xiaochan Wang, and Lei Wang

Abstract Centrifugal variable-rate fertilizer spreaders have attracted attention for their low cost, high efficiency, and reduced crop damage. To improve real-time response and fertilization accuracy, this study developed three actuator-driven feed mechanisms: gear–rack structure A, electric handspike B, and screw slider C. Kinematic models and a time-delay fertilization model were established to analyze lag effects. Performance tests calibrated fertilizer lag distances, actuator response times, and flow rate errors. Results showed that average lag distances for A, B, and C were 3.56 m, 1.85 m, and 2.72 m, with correction times of 1.99 s, 1.48 s, and 1.74 s, respectively. A positive correlation was found between application increment and actuator response time (TBi > TCi > TAi). Flow errors decreased then increased with feed rate, with minimum error between 250–380 g/s, and γBi > γAi > γCi. Structure C was selected for integration, and with a correction time of 1.74 s applied, the average fertilization error reduced to 9.67% and lag distance to 0.37 m. Spreading tests under ISO 5690 and ASAE S341.2 showed optimal uniformity at 300 g/s and 600 r/min, with coefficients of variation of 14.25% and 13.86%. An S-shaped variable-rate model achieved fitting accuracy with R^2 values of 0.91 and 0.82 for increasing and decreasing rates, and a prediction error of 9.47%.

D. Wang · X. Huang · L. Xu
College of Engineering, Nanjing Agricultural University, Nanjing, China
e-mail: 2024212004@stu.njau.edu.cn; 2022212011@stu.njau.edu.cn; 2024812087@stu.njau.edu.cn

Z. Zhao (✉)
College of Information and Electrical Engineering, China Agricultural University, Beijing, China
e-mail: zhaozhangcau@cau.edu.cn

X. Wang · L. Wang
College of Engineering, Nanjing Agricultural University, Nanjing, China

Jiangsu Province Engineering Lab for Modern Facility Agriculture Technology & Equipment, Nanjing, China
e-mail: wangxiaochan@njau.edu.cn; leiwang@njau.edu.cn

Y. Shi et al. (eds.), *Technological Progress on Variable Rate Fertilization*, Smart Agriculture 17, https://doi.org/10.1007/978-981-95-4330-4_6

This study provides a foundation for improving response and uniformity in centrifugal variable-rate fertilization systems.

Keywords Variable-rate fertilization · Feed actuator · Lag correction · Centrifugal spreader · System response

6.1 Introduction

Rational use of fertilizers and pesticides is essential for enhancing rice yield and optimizing harvesting efficiency. In this context, variable-rate technology (VRT) has become an effective approach for precise nutrient application [1–3]. This technology not only meets the growth requirements of rice crops but also supports the improvement of the agroecological environment [4], making it a crucial component of precision agriculture [5, 6]. Variable-rate fertilization for rice involves several key elements, including information acquisition and communication, machinery design, and control system implementation. Among these, the control system serves as the core of VRT, and ensuring accurate data collection along with rapid system response is critical to achieving the intended outcomes of variable-rate fertilization [7–9].

In recent years, researchers worldwide have concentrated on enhancing control systems in variable-rate fertilizer applicators, achieving notable progress in system response performance [9–13]. MALEKI investigated the response delay in a developed variable-rate fertilizer applicator using soil nutrient data [14]. Through static and dynamic experiments, the delay from data acquisition to fertilizer adjustment was analyzed, allowing determination of the optimal sensor placement and system delay correction. Meng Zhijun et al. designed a control system based on a rate prescription map, converting the fertilizer positioning lag into system delay time, and field tests indicated a correction delay of 1.84 s [15]. Qiu Bai-jing et al. developed software for a variable spray device, monitoring real-time response and step characteristics to assess how system delay affects variable spray performance [16, 17]. While these theoretical and experimental studies have improved the operational efficiency of variable-rate fertilization, detailed investigations into control techniques for variable and uniform fertilization using centrifugal fertilizer spreaders equipped with spectral sensors in paddy fields remain limited [18–20].

Building on previous research conducted by our team on variable-rate fertilization, this study focuses on the control system of the developed centrifugal surface variable-rate fertilizer spreader, utilizing real-time rice growth information. The following sections present the development of motion models for three distinct actuators responsible for adjusting application rates, based on structural analysis of the designed adjustment device [21, 22]. Additionally, the fertilization lag model and control system delay response are examined. Performance tests are used to comparatively optimize the fertilizer distribution actuators, and a method for correcting control system response lag is proposed. These results are expected to enhance both the response speed and the distribution accuracy of variable-rate fertilization in rice

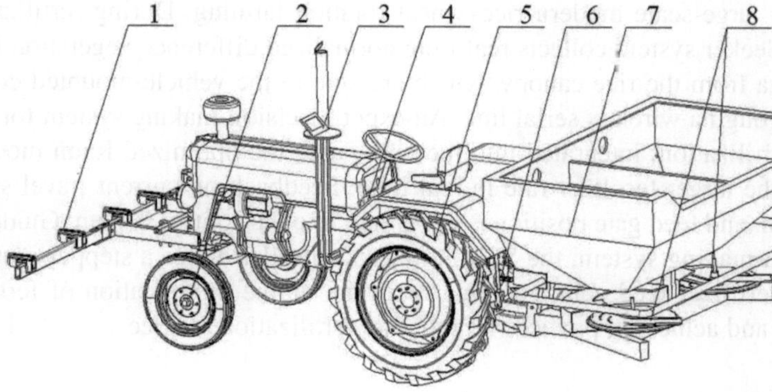

1. Spectral sensor; 2. Speed sensor; 3. On-board CPU; 4. Control system; 5. Driving system;
6. Transmission system; 7. Fertilizer box; 8. Spreader disc

Fig. 6.1 Structural diagram of centrifugal variable-rate fertilizer spreader. 1. Spectral sensor; 2. Speed sensor; 3. On-board CPU; 4. Control system; 5. Driving system; 6. Transmission system; 7. Fertilizer box; 8. Spreader disc

cultivation, thereby improving the operational efficiency of the centrifugal surface variable-rate fertilizer spreader in paddy fields [23–25].

6.2 Machine Structure and Working Principle of the Variable Fertilizer Spreader

6.2.1 Spreader Structure

Figure 6.1 illustrates the overall configuration of the real-time variable-rate fertilizer spreader designed for paddy fields. The system adopts spectral technology and employs dual centrifugal discs. Its main components include a spectral sensing unit, a wireless communication module, a vehicle-mounted control terminal, the fertilizer spreading mechanism, and the control unit, together with several auxiliary parts.

6.2.2 Working Principle

The designed variable-rate fertilizer spreader for rice features a suspended traction structure, high automation, and stable fertilizer distribution performance. The spreading discs are powered in reverse by the tractor's power take-off (PTO), achieving an effective swath width of over 24 m per pass. The disc rotation speed and fertilizer feed gate opening are regulated through a closed-loop feedback system, enhancing fertilizer utilization and production efficiency, making the spreader suitable for large-scale modern rice–wheat rotation farming. During fertilization, the Green Seeker system collects real-time normalized difference vegetation index (NDVI) data from the rice canopy, which are sent to the vehicle-mounted control terminal through a wireless serial link. An expert decision-making system for variable-rate fertilization, integrated into the CPU, uses the optimized Raun model to determine the target fertilizer rate in real time. Feedback on current travel speed, disc rotation, and feed gate position is integrated into the control system. Guided by the decision-making system, the STM32 microcontroller drives a stepper motor to adjust the fertilizer feed gate, enabling real-time online modification of fertilizer application and achieving precise variable-rate fertilization for rice.

6.2.3 Structural Design and Motion Analysis of the Fertilizer Distributing Device

The application rate adjustment device is a key element of the variable-rate fertilizer spreader, as its feeder performance directly affects fertilization accuracy. Based on the operational principles and distributor characteristics of the rice fertilizer spreader, three actuator-based adjustment mechanisms were developed to meet functional requirements. These mechanisms include gear and rack structure A, electric handspike structure B, and screw slider structure C, all controlled by stepper motors.

6.2.3.1 Adjustment Device with Gear and Rack Structure Actuator

Figure 6.2a illustrates the adjustment device with the gear and rack structure actuator, composed mainly of a stepping motor, gear and rack mechanism, connecting rod, moving and fixed disks, angular displacement sensor, and other minor components. Based on the target fertilizer application data provided by the lower computer's decision-making system, the relative position of the fertilizer feed gate, determined by the moving and fixed disks, is adjusted via the gear and rack actuator driven by the stepping motor. This setup enables online control of the fertilizer feed rate within the range of 0 to 500 g/s.

According to theoretical mechanics, the adjustment device was simplified into an offset slider–crank mechanism for motion analysis, enabling determination of the relative rotation angle of the moving disk. Based on this, the corresponding fertilizer flow rate was accurately calculated. The vector model of the mechanism is illustrated in Fig. 6.2b, where R denotes the displacement vector of the slider, $\boldsymbol{Re}$ the eccentricity vector, $\boldsymbol{R_1}$ the crank vector with rotation angle α, and $\boldsymbol{R_2}$ the link vector with rotation angle β. Their magnitudes are represented as r, r_e, r_1, and r_2, respectively. The vector equation is expressed as follows:

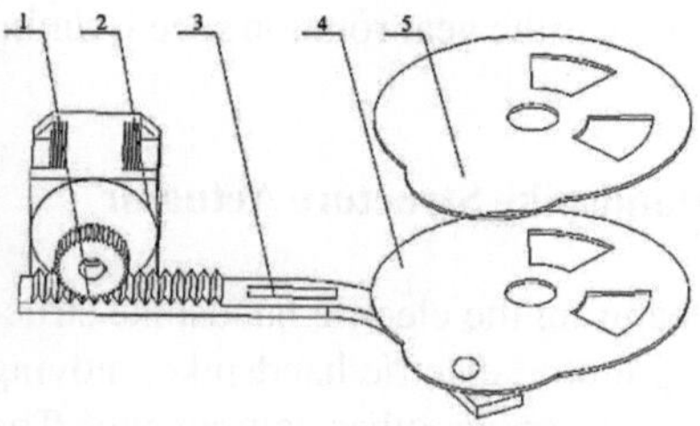

a.Schematic diagram of the adjustment deviceb.

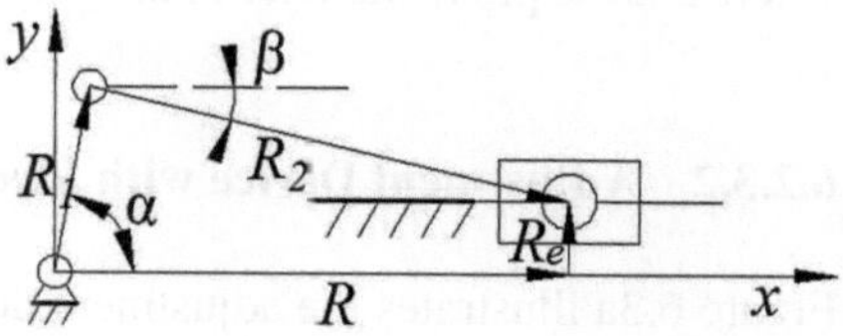

b.Structural diagram of the adjustment device

1. Pinion-rack mechanism; 2. Stepper motor; 3. Link mechanism; 4. Moving disk; 5. Fixed disk

Fig. 6.2 Structural diagram of centrifugal disc and fertilizer regulating device. (**a**) Schematic diagram of the adjustment device. (**b**) Structural diagram of the adjustment device. 1. Pinion-rack mechanism; 2. Stepper motor; 3. Link mechanism; 4. Moving disk; 5. Fixed disk

$$\boldsymbol{R}_1 + \boldsymbol{R}_2 = \boldsymbol{R} + \boldsymbol{R}_e \tag{6.1}$$

Equation (6.1) was decomposed into scalar forms in the x and y axes as follows:

$$\begin{cases} r_1 \cos\alpha + r_2 \cos\beta = r \\ r_1 \sin\alpha + r_2 \sin\beta = r_e \end{cases} \tag{6.2}$$

The velocity equation can be obtained by deriving equation (6.2) with respect to time t:

$$\begin{cases} \dot{r} + r_2\omega_2 \sin\beta = -r_1\omega_1 \sin\alpha \\ r_2\omega_2 \cos\beta = -r_1\omega_1 \cos\alpha \end{cases} \tag{6.3}$$

where, $\dot{r}$ is the change rate of $\boldsymbol{R}$, and indicates the speed of the slider movement; ω_1 and ω_2 are the angular velocities of the crank and link, respectively. Here, when the slider velocity $\dot{r}$, rotation angle β, and angular velocity ω_2 of the link are known, the crank rotation angle α and rotation speed ω_1 can be back calculated. Equation (6.3) was transformed into the matrix form and can be expressed as follows:

$$\begin{pmatrix} -r_1\omega_1 \sin\alpha \\ -r_1\omega_1 \cos\alpha \end{pmatrix} = \begin{pmatrix} r_2 \sin\beta & 1 \\ r_2 \cos\beta & 0 \end{pmatrix} \cdot \begin{pmatrix} \omega_2 \\ \dot{r} \end{pmatrix} \tag{6.4}$$

According to the mechanical principle, the rack velocity for the gear-rack transmission mechanism, v, can be written as

$$v = \frac{\pi dn}{60} (mm/s) \tag{6.5}$$

where d is the pitch diameter of the gear, mm; and n is the gear rotation speed, r/min.

6.2.3.2 Adjustment Device with Electric Handspike Structure Actuator

Figure 6.3a illustrates the adjustment device featuring the electric handspike structure actuator, composed primarily of a stepping motor, electric handspike, moving and fixed disks, angular displacement sensor, and various other components. The procedure for adjusting the fertilizer application rate follows the same process outlined in the previous section.

According to some relevant studies, the adjustment device with the electric handspike actuator was simplified by using the offset oscillating guide bar mechanism to complete the motion analysis. The vector model of the mechanism is shown in Fig. 6.3b; here, $\boldsymbol{R}$ is the frame vector (the fixed point); $\boldsymbol{Re}$ is the eccentricity vector, whose rotation angle is θ; $\boldsymbol{R_1}$ is the crank vector, whose rotation angle is α; $\boldsymbol{R_2}$ is the link vector, whose rotation angle is β; and their norm modules are r, r_e, r_1 and r_2, respectively. The vector equation can be written as follows:

$$\boldsymbol{R}_e + \boldsymbol{R}_2 = \boldsymbol{R}_1 + \boldsymbol{R} \tag{6.6}$$

Equation (6.6) can be decomposed into scalar forms in the x and y axes as follows:

$$\begin{cases} r_e \cos\theta + r_2 \cos\beta = r_1 \cos\alpha + r \\ r_e \sin\theta + r_2 \sin\beta = r_1 \sin\alpha \end{cases} \tag{6.7}$$

Because $\boldsymbol{R_e} \perp \boldsymbol{R_2}$, the conversion of $\beta = \theta - 90°$ can be applied and substituted into eq. (6.7) to get

$$\begin{cases} r_e \cos\theta + r_2 \sin\theta = r_1 \cos\alpha + r \\ r_e \sin\theta - r_2 \cos\theta = r_1 \sin\alpha \end{cases} \tag{6.8}$$

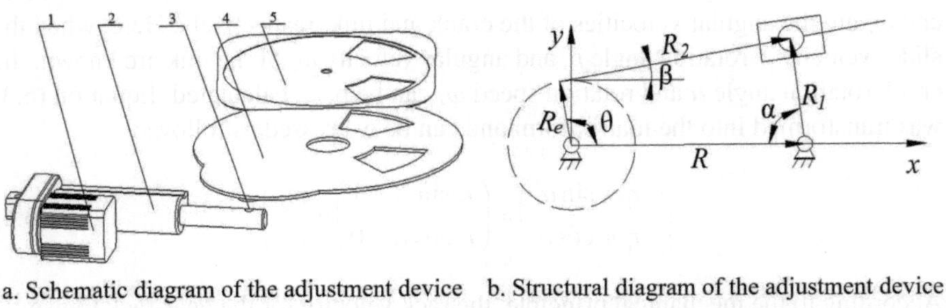

a. Schematic diagram of the adjustment device b. Structural diagram of the adjustment device

1. Stepper motor; 2. Electric handspike; 3. Hinge pin; 4. Moving disk; 5. Fixed disk

Fig. 6.3 Structural diagram of centrifugal disc and fertilizer regulating device. (**a**) Schematic diagram of the adjustment device. (**b**) Structural diagram of the adjustment device. 1. Stepper motor; 2. Electric handspike; 3. Hinge pin; 4. Moving disk; 5. Fixed disk

The velocity equation can be obtained by deriving equation (6.8) with respect to time t:

$$\begin{pmatrix} -r_1\omega_1 \sin\alpha \\ r_1\omega_1 \cos\alpha \end{pmatrix} = \begin{pmatrix} \sin\theta & r_2\cos\theta - r_e\sin\theta \\ -\cos\theta & r_2\sin\theta + r_e\cos\theta \end{pmatrix} \cdot \begin{pmatrix} \dot{r}_2 \\ \omega_2 \end{pmatrix} \tag{6.9}$$

Where, $\dot{r}_2$ denotes the slider movement speed, ω_1 the angular velocity of the crank, and ω_2 the angular velocity of the link. When the slider velocity $\dot{r}_2$, eccentricity rotation angle θ, and link angular velocity ω_2 are known, the crank rotation angle α and angular velocity ω_1 can be back-calculated. In this setup, the slider velocity corresponds to the linear speed of the electric handspike, which can be adjusted by controlling the stepping motor's rotation speed via the program.

6.2.3.3 Adjustment Device with Screw Slider Structure Actuator

Figure 6.4a presents the adjustment device utilizing a screw slider structure actuator, consisting primarily of a stepping motor, screw slider mechanism, moving and fixed disks, lead limit switch, and additional components. The procedure for adjusting the fertilizer application rate follows the same process as described earlier.

In a similar manner, the screw–slider actuator adjustment device was simplified into a planar four-bar tangent mechanism for motion analysis. The vector representation of this mechanism is shown in Fig. 6.4b. Here, $\boldsymbol{R}$ denotes the slider displacement vector; $\boldsymbol{R_0}$ is the distance vector from the fixed point to the screw; $\boldsymbol{R_1}$ is the crank vector with variable length and rotation angle α; and $\boldsymbol{R_2}$ represents the virtual link vector with rotation angle β. Their magnitudes are given as r, r_0, r_1 and r_2, respectively. The corresponding vector equation can be written as

$$\boldsymbol{R}_1 + \boldsymbol{R} = \boldsymbol{R}_2 \tag{6.10}$$

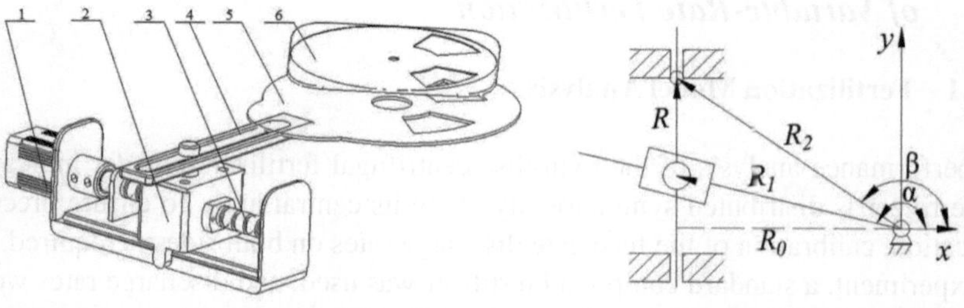

a. Schematic diagram of the adjustment device b. Structural diagram of the adjustment device
1. Stepper motor; 2. Slider; 3. Screw; 4. Chute; 5. Moving disk; 6. Fixed disk

Fig. 6.4 Structural diagram of centrifugal disc and fertilizer regulating device. (**a**) Schematic diagram of the adjustment device. (**b**) Structural diagram of the adjustment device. 1. Stepper motor; 2. Slider; 3. Screw; 4. Chute; 5. Moving disk; 6. Fixed disk

Equation (6.10) can be further expressed in scalar form along the x and y directions as follows:

$$\begin{cases} r_2 \cos(\pi - \beta) = r_1 \cos(\pi - \alpha) \\ r_2 \sin(\pi - \beta) = r + r_0 \tan(\pi - \alpha) \end{cases} \tag{6.11}$$

From Fig. 6.4, it is observed that $r_0 = r_1 \cos(\pi - \alpha)$. By substituting this into Eq. (6.11) and differentiating with respect to time t, we obtain

$$\begin{cases} -\dot{r}_2 \cos\beta + r_2\omega_2 \sin\beta = 0 \\ \dot{r}_2 \sin\beta + r_2\omega_2 \cos\beta = -r_0\omega_1 (\cos\alpha)^{-2} \end{cases} \tag{6.12}$$

where $\dot{r}_2$ is the radial velocity of the link; ω_1 is the angular velocity of the crank; ω_2 is the angular velocity of the link. Equation (6.12) can be transformed into the matrix form and expressed as follows:

$$\begin{pmatrix} 0 \\ -r_0\omega_1 (\cos\alpha)^{-2} \end{pmatrix} = \begin{pmatrix} r_2 \sin\beta & -\cos\beta \\ r_2 \cos\beta & \sin\beta \end{pmatrix} \cdot \begin{pmatrix} \omega_2 \\ \dot{r}_2 \end{pmatrix} \tag{6.13}$$

According to the mechanical principle, the moving distance of the slider in the screw-slider mechanism, L, is

$$L = \frac{nP_h}{60} \tag{6.14}$$

where P_h is the lead of the screw thread, mm; and n is the rotation speed of the screw, r/min.

6.2.4 *System Response Analysis and Mathematical Model of Variable-Rate Fertilization*

6.2.4.1 Fertilization Model Analysis

The performance analysis of the twin-disc centrifugal fertilizer spreader indicates that fertilizer is distributed symmetrically along its central axis. To ensure precise application, calibration of the feed gate discharge rates on both sides is required. In this experiment, a standard compound fertilizer was used, and discharge rates were recorded at gate opening angles of 10°, 15°, 20°, 25°, and 30°. During each test, a collection bag was attached to the feed outlet, and the fertilizer was collected over a 10-second interval. The output rate was determined by weighing the collected fertilizer, with each measurement repeated three times. As shown in Fig. 6.5, the deviation did not exceed 12.14%. Moreover, the flow rate demonstrated a significant quadratic polynomial correlation with the feed gate angle ($P < 0.01$), and the fitted equation is

$$\begin{cases} y_L = 0.6284x^2 - 14.858x + 130.59 \\ y_R = 0.6117x^2 - 14.69x + 128.61 \end{cases} \tag{6.15}$$

The developed variable fertilizer spreader for rice calculates the desired application rate using the expert decision-making system. This system relies on the NDVI values from the crop canopy, which are obtained through the spectral acquisition system, and incorporates the current effective spreading width and feedback travel speed of the machine. Based on this, the mathematical model for the fertilizer application rate per unit area can be expressed as:

$$Q = \frac{2 \cdot q \cdot t}{w \cdot v \cdot t} = \frac{2q}{w \cdot v} \tag{6.16}$$

where Q is the fertilizer application rate per unit area, g/m²; q is the target fertilizer distribution flow rate of a single feed gate, g/s; w is the effective spreading swath width, m; and v is the current travel speed of the entire spreader, m/s.

6.2.4.2 System Response Analysis

After receiving the monitoring data from each sensor, the control system of the developed variable fertilizer spreader directs the expert decision-making system, which then instructs the core controller to adjust the actuator and regulate the fertilizer distribution flow rate. During this process, delays occur due to differences in the

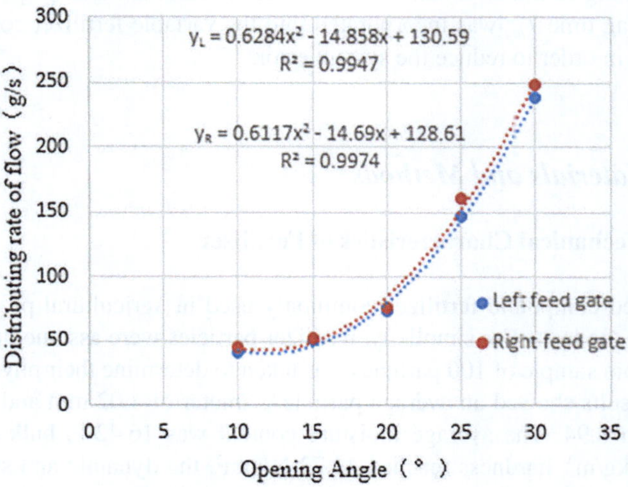

Fig. 6.5 Calibration of particle flow rate

control system and the structural design of the actuator, leading to a lag in the position of fertilizer particles and a response delay in the controller, which impacts fertilization accuracy. Based on the working principle of the spreader, the response lag model for the fertilizer machine system can be described as follows:

$$L = L_1 + L_2 - L_3 \tag{6.17}$$

Where L represents the lag distance of the spreader in dropping fertilizer (m), L_1 is the distance from the illuminant of the spectral acquisition system to the fertilizer feed gate of the spreader (m), L_2 is the distance from the exit of the spreading disc to the point where the fertilizer particles land on the ground (m), and L_3 is the distance traveled by the spreader from the point where spectral data is acquired to where the fertilizer particles fall on the ground (m).

As shown in Fig. 6.1 and Eq. (6.17), L_1 is dependent on the installation position of the spectral sensors, while L_2 is determined by the structural design of the fertilizer spreader, and its exact value can be obtained through theoretical calculations. L_3, on the other hand, is influenced by several factors such as the walking speed of the spreader, the response time of the control system, and the execution time of the mechanism. Since L_3 is a composite value that affects decision-making and is challenging to measure directly, it can be calibrated during testing. Dynamic adjustments can then be made to minimize fertilization lag. The conversion formula relating the system's time delay to the fertilizer lag distance can be expressed as follows:

$$T_{lag} = \frac{L}{v} \tag{6.18}$$

Here, T_{lag} is the spreader system lag time, s; L is the gauged lag distance of the spreader falling fertilizer, m; and v is the spreader current walking speed, m/s. The correction lag time T_{lag} was incorporated into the variable fertilizer control system in advance, in order to reduce the system error.

6.2.5 Materials and Methods

6.2.5.1 Mechanical Characteristics of Fertilizer

A KCl-based compound fertilizer commonly used in agricultural production was selected for the tests. For simplicity, fertilizer particles were assumed to be spherical. A random sample of 100 particles was taken to determine their physical properties. The results showed an average particle diameter of 4.02 mm and a sphericity greater than 0.94. The average moisture content was 16.42%, bulk density was 1.38×10^3 kg/m^3, hardness reached 44.73 N/mm^2, the dynamic and static friction coefficients were 0.26 and 0.34, respectively, Poisson's ratio was 0.24, and the shear modulus was 10.76 MPa.

6.2.5.2 Performance Test Method

Performance tests were conducted in April 2017 at the Yancheng Yanhai Tractor Manufacturing Co., Ltd. test station in Jiangsu. The conditions were favorable: clear weather, temperature between 15 and 25 °C, wind speed below 2.0 m/s (meeting ASAE test requirements), flat terrain, a test area of 450 m^2, relative air humidity of 42%, and soil moisture content of 21%. Test methods and evaluation indices followed the relevant standards, including GB/T5262-2008 (General rules for the determination of agricultural mechanical test conditions), ISO 5690, and ASAE S314.2. The experimental setup is illustrated in Fig. 6.6.

The test area, measuring 30 m × 14 m, was covered with plastic film to minimize fertilizer loss, and soft cotton cloth was placed inside each collection pan to reduce errors caused by fertilizer bouncing. A total of 150 collection pans, each sized 50.8 cm × 40.6 cm × 10.2 cm, were arranged in 15 rows and 10 columns. The spacing between the pans was 1.5 m in the columns and 0.8 m in the rows. The amount of fertilizer in each pan was measured to analyze the transverse distribution of fertilizer particles, as the spreader moved across the center of the test area at a speed of 1.2 m/s. The bulk of fertilizer in the spreader was adjusted to approximately 50% of the pan's total capacity, with calibration conducted to match the working conditions for each test. After stabilizing in the designated stable region, the spreader moved through the measurement area (the section with the collection pans). A stop area was included at the end to ensure the test's accuracy. The coefficient of variation for the transverse distribution of fertilizer particles was calculated to evaluate the spreading performance of the centrifugal variable-rate fertilizer applicator, and each set of experiments was repeated three times. The walking path and the arrangement of the collection pans are shown in Fig. 6.7, with the equations for the coefficient of variation provided below:

Fig. 6.6 Performance test of spreading fertilizer

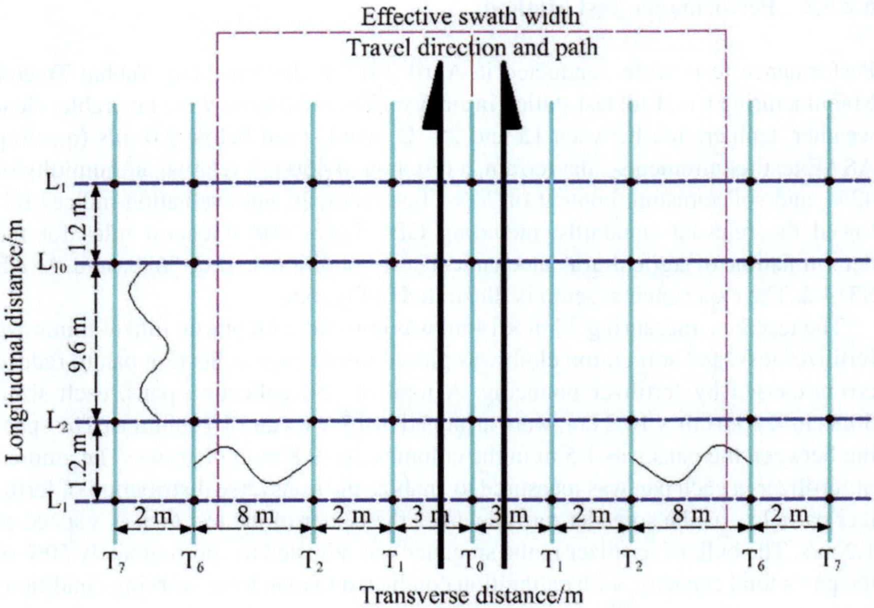

Fig. 6.7 Schematic diagram of collection pan matrix

$$\begin{cases} CV = \dfrac{SD}{\overline{X}} \times 100\% \\ SD = \left\{ \sum_{i=1}^{n} \left[\left(X_i - \overline{X} \right)^2 \right] / \left(n-1 \right) \right\}^{\frac{1}{2}} \\ \overline{X} = \dfrac{1}{n} \sum_{i=1}^{n} X_i \qquad \left(i = 1,2,3,\ldots,n \right) \end{cases} \tag{6.19}$$

where, *CV* is the coefficient of variation of particle distribution; *SD* is the standard deviation of fertilizer quantity; $\overline{X}$ is the average quantity of fertilizer particles; *Xi* is the quantity of fertilizer particles in the *i*th collection pan g; and *n* is the total number of collection pans.

6.2.5.3 Field Test Method

In June 2017, field validation tests were conducted at a rice test base at Dongtai Jinmansui Agricultural Development Co., Ltd. in Jiangsu to verify the spreading performance and prediction accuracy of the variable fertilization model developed in for both increasing and decreasing distribution flow rates. The weather was clear, with a temperature between 25 °C–36 °C, the wind speed was under 2.0 m/s, the testing area was 20 acres, the relative humidity of the air was 31%, and the test was conducted in accordance with the methods detailed in Sect. 6.2.5.2.

6.3 Results and Analysis

6.3.1 Execution System Response Tests of the Variable Fertilizer Spreader

To evaluate the time response of the variable fertilizer control system, tests were conducted using three different fertilizer distributing actuators: gear and rack structure (*A*), electric handspike structure (*B*), and screw slider structure (*C*). The system's response times were assessed by calibrating the lag distances of fertilizer falling from the spreader, and the accuracy of the fertilizer distribution rate for each actuator was also measured. These tests aimed to determine how each actuator influenced the performance of the variable fertilizer spreader in terms of response delay and distribution precision.

6.3.1.1 Test Conditions and Methods

The calibration tests for the lag distance were conducted at a rice test base in Dongtai Jinmansui Agricultural Development Co. Ltd., Jiangsu, China, in July 2017. The weather conditions were clear, with temperatures ranging from 27–38 °C and relative humidity at 31%. The test used a compound granular fertilizer, widely utilized in southern China, developed by Nanjing Zhengmeishi Agrochemical Co. Ltd. The fertilizer had a moisture content of 1.03%, and its mean particle diameter was 4.02 mm. The fertilizer spreader was mounted on a DF1204 (Dong Feng) tractor using a three-point suspension system. The test layout for the system response is shown in Fig. 6.8.

During the experiments, the fertilizer spreader equipped with three actuator types (*A*, *B*, and *C*) was driven across a 5 m × 50 m test plot at a controlled walking speed. The start and end lines were marked in red and blue, respectively. When the spectral sensor detected the red boundary (point 1 in Fig. 6.8), the test began. At this moment, the ground speed (v) was measured with Hall sensors, and the stopwatch was triggered as the distributing actuator initiated fertilizer discharge (point 2 in Fig. 6.8). The horizontal distance from the red boundary to the initial dropping point was measured with a tape, defined as the fertilizer start lag distance (L_{start}). When the actuator's moving plate rotated to the target position corresponding to a given fertilizer rate (0–500 g/s), the response time T1 was recorded. Similarly, when the spectral sensor scanned the blue boundary (point 3 in Fig. 6.8), the stopwatch continued until the fertilizer stopped falling (point 4 in Fig. 6.8). The horizontal distance between the end point of fertilizer dropping and the blue boundary was measured as

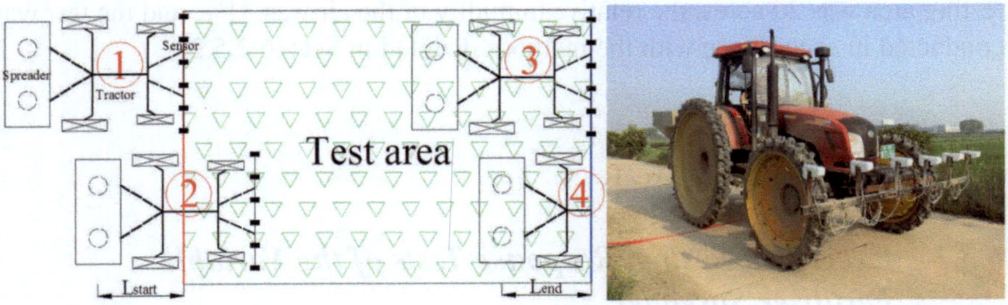

Fig. 6.8 Layout for test of time response of control system

the fertilizer stop lag distance (L_{end}). At the same time, the response time T_2 was recorded when the actuator returned to the off state (fertilizer flow rate = 0 g/s). The discharged fertilizer mass (*M*) was weighed using an electronic scale. Each test was repeated three times, and average values were calculated. To ensure accuracy, the fertilizer cans were kept above 50% of their total capacity, and the spreader was calibrated to its working state before each run.

6.3.1.2 Test Results and Analysis

The calibration results for fertilizer falling lag distance with the three actuators are presented in Fig. 6.9. Distinct differences were observed among actuators, with average lag distances ranked as $L_A > L_C > L_B$, corresponding to 3.56 m, 2.72 m, and 1.85 m, respectively. The maximum lag distance of 3.91 m occurred with actuator A at a walking speed of 1.83 m/s, while the minimum lag distance of 1.42 m occurred with actuator B at 1.48 m/s. As shown in Fig. 6.9, the fertilizer falling lag distance increases proportionally with the walking speed of the spreader. Based on the average lag distances and walking speeds, the fertilizer falling delay correction times were calculated as 1.99 s, 1.48 s, and 1.74 s for actuators *A*, *B*, and *C*, respectively.

After the fertilizer spreader's control system received the command for a specified fertilizer application rate, the time required for the actuators to adjust the feed gate plate to the desired opening was evaluated. The response times of the three actuators (*A*, *B*, *C*) for different target application rates were recorded and are shown in Fig. 6.11. As the target fertilizer application rate increased from 100 to 500 g/s, the response time for fertilization gradually increased. For actuators *A*, *B*, and *C*, the minimum and maximum mean response times were as follows: 0.86 s to 3.53 s for *A*, 1.55 s to 3.78 s for *B*, and 1.06 s to 3.76 s for *C* (with the minimum and maximum corresponding to 100 g/s and 500 g/s, respectively). When comparing the response times for the three actuators, the results show that $T_{Bi} > T_{Ci} > T_{Ai}$, where i indicates the corresponding fertilizer flow rate. The opening and closing response times are considered as the delays in response when the fertilizer application rate is either increasing or decreasing to the target value.

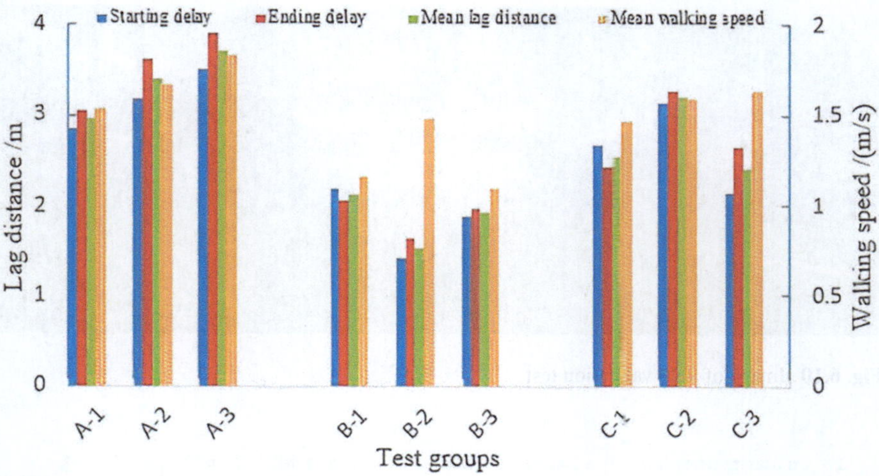

Fig. 6.9 Calibration test results of delay distance

6.3.1.3 Field Tests and Observations

The developed centrifugal variable fertilizer spreader, equipped with the selected adjustment device (actuator C) and a fertilizer cover for rice, was field-tested at the Dongtai test base. To verify the system's time response and fertilizer distribution accuracy, the modified response time for fertilizer lag delay (T_{lag} = 1.74 s) was incorporated into the control system. This approach aligns with the methodology discussed earlier in this section. The actual field test site is depicted in Fig. 6.10.

Table 6.1 summarizes the results of the field fertilizer tests conducted with the control system, which was modified to include the corrected lag time in the variable fertilizer spreader. Out of six test groups, the mean error in fertilizer distribution was 9.67%. The maximum lag distance of falling fertilizer was 0.82 m, while the minimum lag distance was 0.19 m, and the average lag distance was 0.37 m. Notably, there was a phenomenon where fertilizer fell ahead of the spreader. These results suggest that the correction of system response has effectively reduced the delay in fertilizer distribution, and actuator *C*, in particular, showed high accuracy in fertilizer application.

As the actuator adjusted the feed gate plate of fertilizer cans to the corresponding target fertilizer application rate, the error between the actual distributing fertilizer application rate and target fertilizer application rate will remain, and the test results of distributing fertilizer accuracy for the three actuators (*A*, *B*, *C*) within a certain period of time were shown in Fig. 6.12. The maximum error in the fertilizer distribution flow rate is 17.82%, which occurs for actuator *B*; the target fertilizer application rate is 100 g/s, and the corresponding average operating speed of the spreader is 1.68 m/s. The minimum error is 8.24%, which occurs for actuator *C*; the target fertilizer application rate is 300 g/s, and the corresponding average operating speed is 1.15 m/s. For actuators A, B, and C, the maximum and minimum mean errors are

Fig. 6.10 Image of field validation test

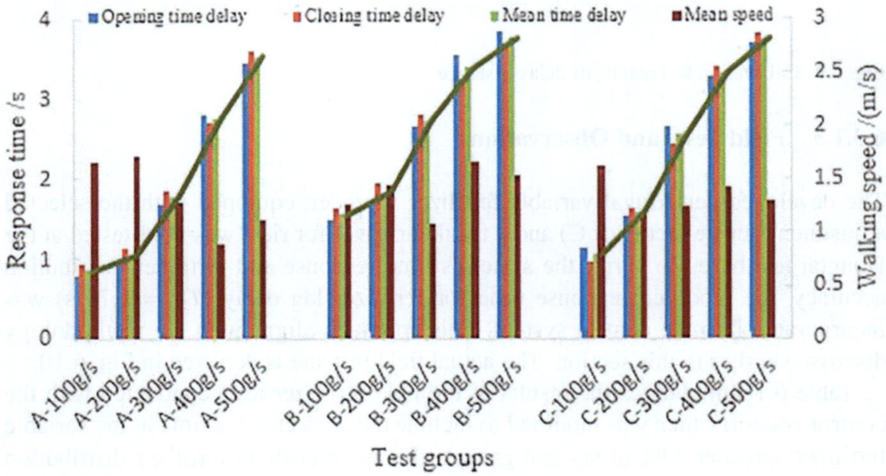

Fig. 6.11 Calibration test results of response delay time

Table 6.1 Results of field test

Test	Lag distance of fertilizer falling (m)				Fertilizer distribution accuracy (%)
	Starting lag	Ending lag	Mean lag	Mean speed (m/s)	
1	0.74	0.37	0.56	1.18	12.21
2	0.82	−0.42	0.2	1.36	6.82
3	0.56	0.68	0.62	1.04	9.47
4	0.44	0.19	0.32	1.23	11.36
5	−0.27	0.46	0.1	1.57	10.52
6	0.31	0.52	0.41	1.26	7.63
Mean value	/		0.37	1.27	9.67

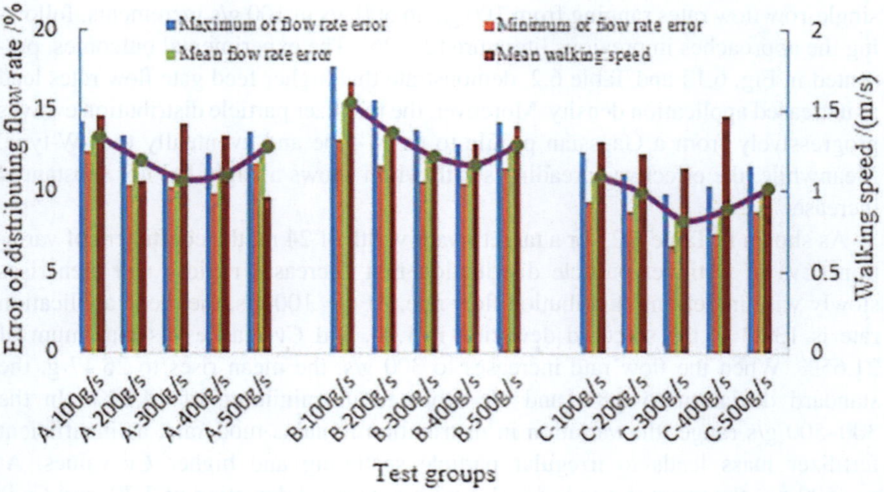

Fig. 6.12 Test results of the accuracy of application rate

13.41% and 10.74%; 15.54% and 11.71%; and 10.89% and 8.24%, respectively. The respective average error values are 11.97%, 13.19%; and 9.62%. A comparison of the effect of the actuators on the accuracy of fertilizer distribution shows that $\gamma_{Bi} > \gamma_{Ai} > \gamma_{Ci}$. The accuracy first increases and then decreases in the range of target fertilizer application rate of 100–500 g/s; this indicates that the error of the fertilizer distribution flow rate first decreases and then increases. The extremum appears between 250 and 380 g/s.

Based on the analysis, the performance of the different actuators (*A*, *B*, and *C*) in the fertilizer control system was evaluated. Actuator *B* exhibited the shortest lag distance and the least correction for lag time, followed by actuators *C* and *A*. Actuator *A* demonstrated the fastest response time, with actuators *C* and *B* coming next. In terms of fertilizer distribution accuracy, actuator *C* outperformed the others, showing the least error, followed by actuators *A* and *B*. Therefore, actuator *C*, which is based on the screw slider structure, was chosen as the fertilizer adjustment device for the variable fertilizer spreader in paddy fields. This actuator provides a shorter lag distance, faster executive response time, and superior fertilizer distribution accuracy.

6.3.2 *Fertilization Strategy Using Feed Gate Flow Rate*

When operating the developed variable-rate fertilizer spreader, the application density is primarily regulated by adjusting the distributor's feed gate flow rate. To investigate how the flow rate influences the spreading performance of the centrifugal applicator, tests were conducted with a disc speed fixed at ω = 600 r/min and single-row flow rates ranging from 100 g/s to 500 g/s in 100 g/s increments, following the approaches in previous literature [21, 26]. The experimental outcomes, presented in Fig. 6.13 and Table 6.2, demonstrate that higher feed gate flow rates lead to increased application density. Moreover, the fertilizer particle distribution evolves progressively from a Gaussian profile to an M-type and eventually to a W-type. Meanwhile, the effective spreading swath width shows a slight but not substantial increase.

As shown in Table 6.2, for a target swath width of 24 m, the coefficient of variation (*Cv*) of fertilizer particle distribution first decreases rapidly and then rises slowly with increasing distribution flow rate. At q = 100 g/s, the mean application rate is 13.23 g, the standard deviation is 1.80, and *Cv* reaches its maximum of 21.65%. When the flow rate increases to 300 g/s, the mean rises to 26.47 g, the standard deviation to 2.81, and *Cv* drops to its minimum of 14.25%. In the 300–500 g/s range, the variation in distribution remains moderate, as insufficient fertilizer mass leads to irregular particle scattering and higher *Cv* values. At q = 500 g/s, the average rate is 44.21 g with a standard deviation of 3.70, and *Cv* is only 19.44%.Further analysis indicates that higher flow rates enlarge the effective spreading width because particle interactions intensify with increasing application rates. Within this width, the localized application rate slightly exceeds the overall

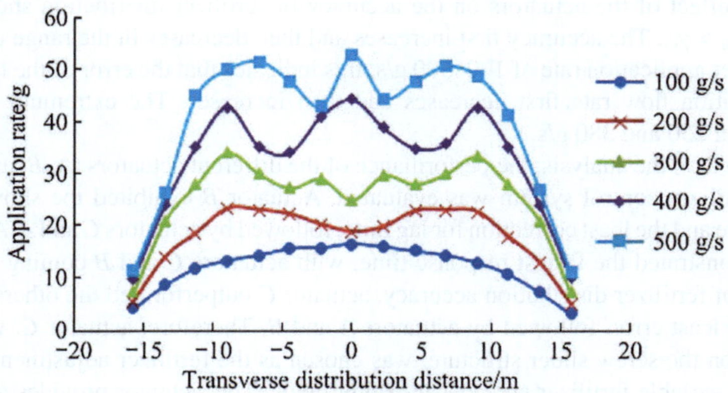

Fig. 6.13 Effects of feed gate flow rate on spreading performance

Table 6.2 Performance test results of flow rate

Flow rate/g·s^{-1}	Target breadth/m	Application density/g			Effective breadth/m	Application density/g		
		Mean/g	SD	Cv/%		Mean/g	SD	Cv/%
100	24	13.23	1.80	21.65	19.64	14.87	1.19	10.05
200		19.88	2.81	17.24	21.52	20.95	1.67	12.63
300		26.47	2.63	14.25	24.06	29.19	2.84	13.89
400		35.91	3.83	17.82	25.37	38.11	3.97	12.37
500		44.21	3.70	19.44	27.41	47.45	3.25	15.26

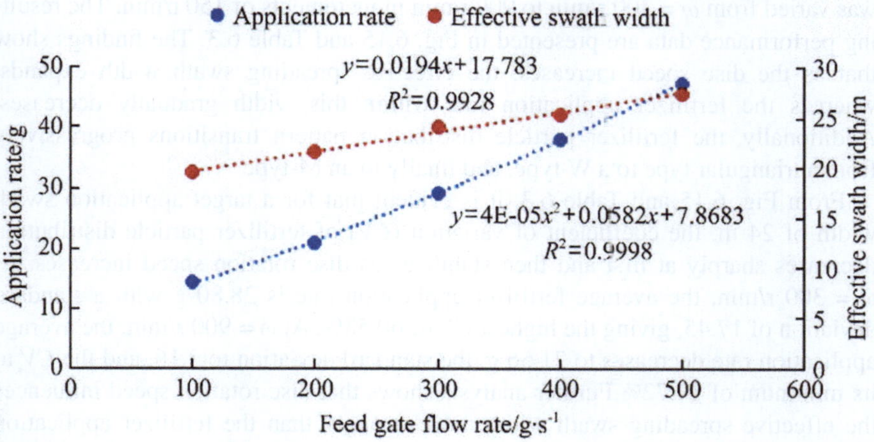

Fig. 6.14 Relationship between feed gate flow rate and application rate, effective swath width

average. For example, at q = 100 g/s, the mean rate is 14.87 g, and *Cv* reaches its lowest value of 10.05%, reflecting the Gaussian-type advantage. Across all feed gate flow rates, *Cv* values remain below 15.26%, meeting the requirement of $Cv \leq 20\%$ specified in NYT1003-2006, Technical specification of quality evaluation for fertilizing machinery.

To evaluate how distribution flow rate influences the performance parameters of the spreader, regression analyses were carried out between the experimentally controlled feed gate flow rate and two key metrics: the measured fertilizer application rate and the effective spreading swath width. The fitted equations and their coefficients of determination are presented in Fig. 6.14 ($p < 0.01$). Results indicate that, across the 100–500 g/s range, the fertilizer application rate follows a quadratic polynomial relationship with the distribution flow rate, whereas the effective swath width exhibits a linear dependence. Both show strong positive correlations. The corresponding regression models for application density (y_{rate}) and spreading swath width (y_{width}) are expressed in Eq. (6.20):

$$\begin{cases} y_{rate} = 4E-05x^2 + 0.0582x + 7.8683 \\ y_{width} = 0.0194x + 17.783 \end{cases} \tag{6.20}$$

6.3.3 *Fertilization Strategy Using Disc Rotation Speed*

The spreading swath width of the developed centrifugal variable-rate spreader can be regulated by adjusting the disc rotation speed. To investigate this effect, the same experimental procedures used for evaluating feed gate flow rate were applied. In this test, the distribution flow rate was fixed at $q = 300$ g/s, while the disc rotation speed was varied from $\omega = 300$ r/min to 900 r/min in increments of 150 r/min. The resulting performance data are presented in Fig. 6.15 and Table 6.3. The findings show that as the disc speed increases, the effective spreading swath width expands, whereas the fertilizer application rate within this width gradually decreases. Additionally, the fertilizer particle distribution pattern transitions progressively from a triangular type to a W-type, and finally to an M-type.

From Fig. 6.15 and Table 6.3, it is evident that for a target application swath width of 24 m, the coefficient of variation (*CV*) of fertilizer particle distribution decreases sharply at first and then stabilizes as disc rotation speed increases. At $\omega = 300$ r/min, the average fertilizer application rate is 28.80 g with a standard deviation of 17.45, giving the highest *CV* of 60.58%. At $\omega = 900$ r/min, the average application rate decreases to 21.56 g, the standard deviation to 3.16, and the *CV* to its minimum of 14.72%.Further analysis shows that disc rotation speed influences the effective spreading swath width more strongly than the fertilizer application density. At $\omega = 300$ r/min, the effective swath width is at its minimum of 12.67 m, with an average application rate of 42.62 g, a standard deviation of 9.09, and a *CV* of 21.34%. At $\omega = 900$ r/min, the effective swath width reaches its maximum of 28.09 m, while the average application rate is 25.18 g, with a standard deviation of

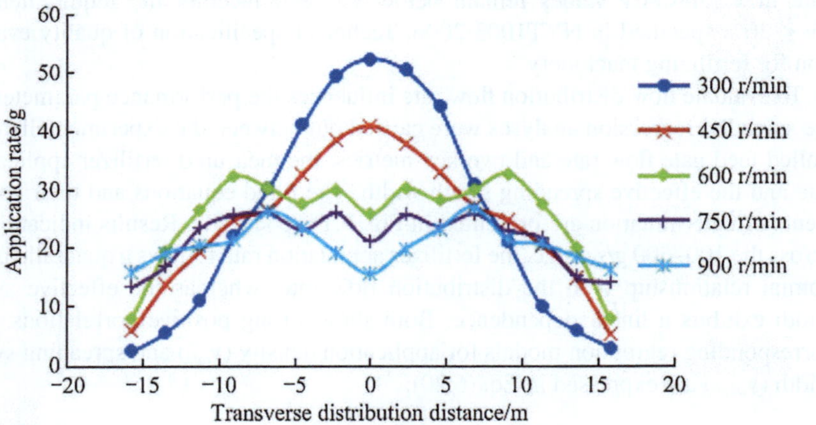

Fig. 6.15 Relationship between disc rotation speed and application rate, effective swath width

Table 6.3 Performance test results of disc rotation speed

Disc speed/r·min^{-1}	Effective breadth/m	Application density/g			Target breadth/m	Application density/g		
		Mean/g	SD	Cv/%		Mean/g	SD	Cv/%
300	12.67	42.62	9.09	21.34	24	28.80	17.45	60.58
450	17.84	33.24	6.26	19.66		29.67	8.86	32.43
600	24.51	28.36	2.23	13.83		27.66	4.07	14.97
750	25.73	26.53	3.48	15.35		24.15	3.48	15.19
900	28.09	25.18	3.16	14.42		21.56	3.16	14.72

3.16 and a *CV* of 14.42%. Notably, the application rate within the effective swath width is generally higher than that over the full spreading width, which can be attributed to the increase in centrifugal force and particle velocity with higher disc speeds. This enlarges the spreading width and produces a more uniform particle distribution, reducing fertilizer concentration in localized areas under constant flow rate conditions.When the disc speed exceeds 600 r/min, the distribution pattern shows little further variation, *CV* remains below 15.26%, and the swath width tends to stabilize. However, at high rotation speeds, fertilizer particle stratification becomes apparent and power consumption rises. Therefore, ω = 600 r/min is identified as the critical speed, yielding an average effective swath width of 24.51 m, an average application rate of 28.36 g, a standard deviation of 2.23, and the lowest *CV* of 13.83%, indicating optimal spreading uniformity.

To quantify the influence of disc rotation speed on spreader performance, regression analyses were conducted between disc speed and both effective spreading swath width and fertilizer application rate. The resulting fitting equations and coefficients of determination are provided in Fig. 6.16 ($p < 0.05$). The results show that within the range of 300–900 r/min, effective swath width exhibits a positive quadratic polynomial relationship with disc speed, whereas fertilizer application rate follows a negative power correlation. The corresponding fitting models for effective swath width (y_{width}) and application rate (y_{rate}) are expressed in Eq. (6.21).

$$\begin{cases} y_{width} = -4E-05x^2 + 0.0725x - 5.994 \\ y_{rate} = 1615.4x^{-0.638} \end{cases} \tag{6.21}$$

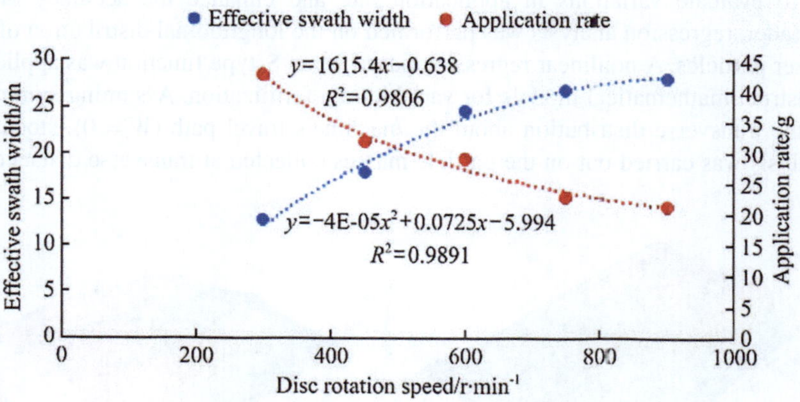

Fig. 6.16 Effect of disc rotation speed on spreading performance

6.3.4 Fertilization Strategy Using Variable Control

In the developed centrifugal variable-rate fertilizer spreader, the fertilizer application rate is dynamically adjusted by varying the feed gate opening according to crop growth information acquired in real time from the spectral sensor. To analyze fertilizer particle distribution during variable-rate operation and assess spreader performance, a field test was conducted with the spreader ground speed (v) set at 1.2 m/s, the disc rotation speed (ω) fixed at 600 r/min, and the distribution flow rate (q) gradually increased from 100 g/s to 500 g/s, then decreased back to 100 g/s. The performance results of this test are presented in Fig. 6.17. For reference, Fig. 6.17 designates the longitudinal axis position of 0 m as the start of variable-rate fertilization (steady distribution rate of q = 100 g/s) and 40 m as the end of fertilization (steady distribution rate of q = 500 g/s).

It can be seen from Fig. 6.17 that as the distribution flow rate increases from 100 g/s to 500 g/s, the fertilizer particle pattern evolves sequentially from Gaussian-type to M-type and finally to W-type, consistent with the results presented in Sect. 4.1. When the flow rate decreases from 500 g/s back to 100 g/s, a similar transformation occurs but with distinct transition characteristics. These differences arise from both the control behavior of the distribution actuator and the greater dispersibility of fertilizer particles. In addition, as the application rate rises during variable fertilization, the effective swath width expands. To ensure uniformity under multi-trajectory operation of the centrifugal spreader, it is necessary to constrain variable-rate fertilization to a single trajectory, thereby minimizing excessive swath widening and enhancing distribution quality within the target width.

To evaluate variations in application rate and enhance the accuracy of the spreader, regression analysis was performed on the longitudinal distribution of fertilizer particles. A nonlinear regression based on an S-type function was applied to construct mathematical models for variable-rate fertilization. Assuming symmetry of the transverse distribution about the machine's travel path (W = 0), stochastic analysis was carried out on the particle masses collected at transverse distances of

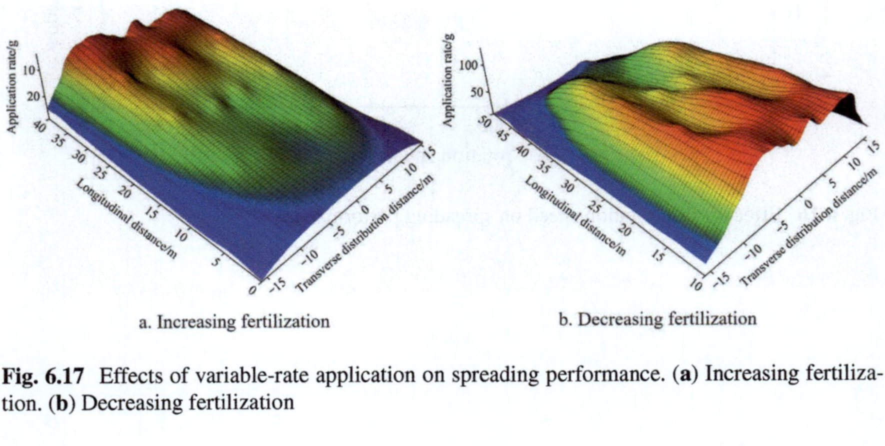

Fig. 6.17 Effects of variable-rate application on spreading performance. (**a**) Increasing fertilization. (**b**) Decreasing fertilization

±2.25 m, ±4.5 m, ±6.75 m, ±9 m, ±11.25 m, ±13.5 m, and ±15.75 m from the centerline. Each test was conducted three times, and mean values were used for fitting. The applied S-type regression function is expressed in Eq. (6.22), while the fitted curves are shown in Fig. 6.18, where different colored lines correspond to the collection boxes at the respective transverse distances.

$$y = y_0 + \frac{a}{1 + e^{-\left(\frac{x - x_0}{b}\right)}} \tag{6.22}$$

where xxx represents the longitudinal distance (m); x_0 is the longitudinal distance at which y reaches half of its maximum value (m); y_0 denotes the terminal application rate (extremum, g); a is the range, defined as the difference between the maximum and minimum values of y (g); and b is a constant.

Table 6.4 summarizes the variable fertilization models obtained from nonlinear regression fitting curves as the distribution flow rate of the developed spreader increased (100–500 g/s) and decreased (500–100 g/s). Combined with Fig. 6.18, the results show that during increasing flow, the S-shaped regression curve provided the best fit, achieving $R^2 = 0.99$ at the central symmetry line ($W = 0$ m). This confirms the high accuracy of the variable fertilization regression model. As the transverse distance widened, fertilizer application rates declined and the fit decreased slightly, though R^2 values remained above 0.82. A comparable trend was observed under decreasing flow: the maximum fit at $W = 0$ m reached $R^2 = 0.86$, with all values exceeding 0.71. Overall, the fit was marginally weaker than in the increasing flow case, likely due to the multilayer deposition pattern created during centrifugal spreading.

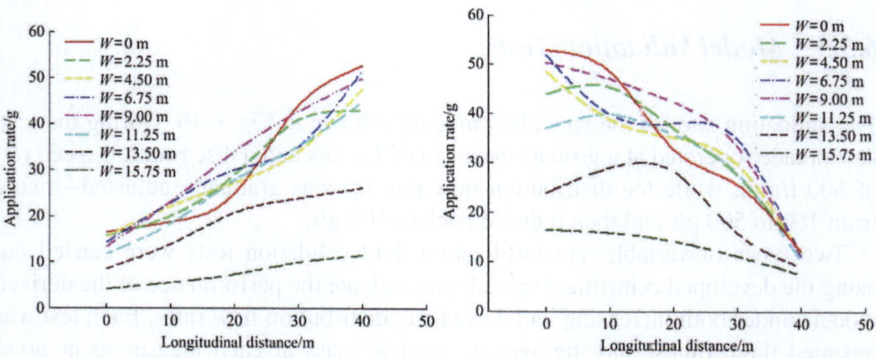

a. Increasing fertilization b. Decreasing fertilization

Fig. 6.18 Regression curves of variable-rate fertilization model. (**a**) Increasing fertilization. (**b**) Decreasing fertilization

Table 6.4 Parameters of variable-rate fertilization model

Parameters		Transverse pan position /m							
		0	±2.25	±4.5	±6.75	±9.0	±11.25	±13.5	±15.75
Increasing	a	37.18	28.16	19.87	88.53	69.91	135.78	22.19	9.37
	b	4.41	5.22	5.35	3.65	16.53	37.36	8.27	9.55
	x_0	25.59	20.13	24.57	34.23	12.83	40.88	12.07	22.42
	y_0	16.89	15.67	18.21	−12.28	−8.71	−22.37	4.37	3.38
	R^2	0.99	0.94	0.88	0.85	0.96	0.93	0.87	0.81
Decreasing	a	37.18	32.98	47.56	97.20	55.52	44.76	24.43	9.74
	b	−4.41	−7.56	−10.42	−5.69	−12.60	−11.88	−8.34	−8.84
	x_0	14.40	18.98	29.68	17.41	25.36	23.00	27.50	21.59
	y_0	16.89	13.75	11.36	4.15	−0.25	2.08	2.80	2.20
	R^2	0.86	0.83	0.72	0.80	0.84	0.75	0.81	0.71

Fig. 6.19 Picture of field validation test

6.3.5 Model Validation Tests

The validation site for the spreading model is shown in Fig. 6.19. During the test, the spreader operated at a ground speed (v) of 1.2 m/s and a disc rotation speed (ω) of 600 r/min, while the distribution flow rate (q) was gradually adjusted—rising from 100 to 500 g/s and then reduced back to 100 g/s.

Two types of variable-rate fertilization field validation tests were carried out using the developed centrifugal spreader to evaluate the performance of the derived models under both increasing and decreasing distribution flow rates. Each test was repeated three times, and the average particle mass at each measurement point within the test area was recorded (results shown in Fig. 6.20). The mean relative error between the model-predicted and experimental fertilizer application rates was 9.47%, with a maximum of 16.62%, a minimum of 4.19%, and a deviation of 4.84%. These findings demonstrate that the predictive accuracy of the variable-rate fertilization model exceeded 80%. The correlation between predicted and measured application rates is presented in Fig. 6.21.

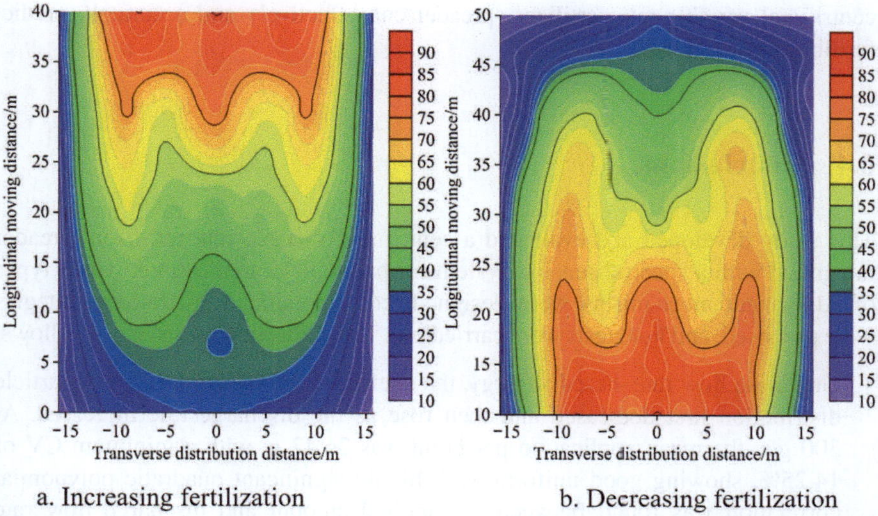

Fig. 6.20 Results of field validation test. (**a**) Increasing fertilization. (**b**) Decreasing fertilization

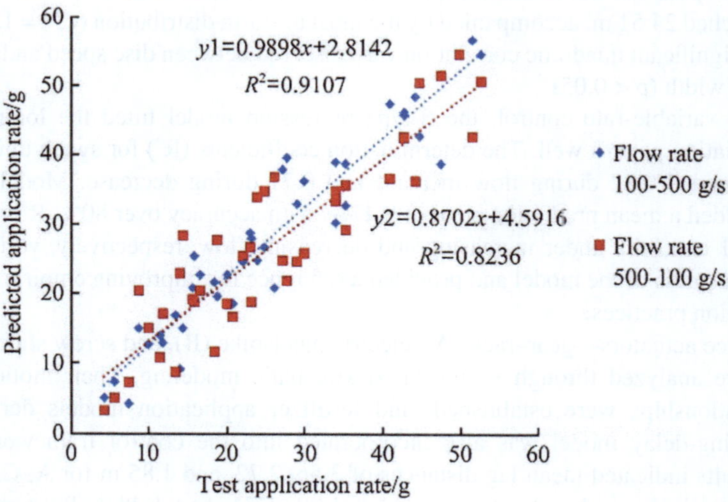

Fig. 6.21 Correlation analysis of test values and predicted values

The correlation analysis between the variable-rate fertilization model and experimental results (Fig. 6.21) shows determination coefficients (R^2) of 0.91 for the increasing rate test and 0.82 for the decreasing rate test, with corresponding correlation line slopes of 0.98 and 0.87. These outcomes demonstrate a strong agreement between model predictions and measured values, confirming the accuracy of the model for variable fertilization and the reliability of the selected reference values. Overall, the field tests validate that the distribution model of the developed centrifugal variable-rate fertilizer spreader can effectively and accurately predict fertilizer application.

6.4 Conclusions

This study developed and evaluated a centrifugal variable-rate fertilizer spreader integrated with a spectral sensing system for paddy field applications. Three types of adjustment mechanisms were designed, corresponding fertilization strategies were proposed, and field tests were carried out. The main conclusions are as follows:

1. Under the flow rate–based strategy, the coefficient of variation (CV) of particle distribution first decreased and then rose as the discharge rate increased. At 300 g/s, the mean application per point was 26.47 g, with a minimum CV of 14.25%, showing good uniformity. A highly significant quadratic polynomial correlation was found between the applied amount and discharge flow rate ($p < 0.01$).
2. With the disc speed–based strategy, the working swath broadened with rising disc speed. At the threshold speed of 600 r/min, the effective spreading width reached 24.51 m, accompanied by the most uniform distribution (CV = 13.83%). A significant quadratic correlation was observed between disc speed and spreading width ($p < 0.05$).
3. For variable-rate control, the S-type regression model fitted the longitudinal variation pattern well. The determination coefficients (R^2) for swath uniformity exceeded 0.82 during flow increase and 0.71 during decrease. Model testing yielded a mean prediction error of 9.47%, with accuracy over 80%. R^2 values of 0.91 and 0.82 under increasing and decreasing flow, respectively, verified the robustness of the model and provided a reference for improving empirical fertilization practices.
4. Three actuators—gear–rack (A), electric handspike (B), and screw slider (C)—were analyzed through vector-based kinematic modeling. Their motion–flow relationships were established, and fertilizer application models derived. A falling-delay model was also incorporated into the control framework. Test results indicated mean lag distances of 3.56, 2.72, and 1.85 m for A, C, and B, respectively, with correction times of 1.99, 1.74, and 1.48 s. Response time increased with target rate, following $T_{Bi} > T_{Ci} > T_{Ai}$. The most stable accuracy was achieved in the 250–380 g/s range, with errors in the order $\gamma_{Bi} > \gamma_{Ai} > \gamma_{Ci}$.
5. Comprehensive evaluation of responsiveness and stability identified actuator C as the best choice. A lag correction time of $T\_lag = 1.74$ s was introduced into the control program. Field verification demonstrated a mean application error of 9.67% and a fertilizer lag distance of 0.37 m, confirming that the optimized control scheme effectively reduced delay and enhanced application precision in paddy fields.

6. External influences—including paddy field surface conditions, machine dynamics, communication latency in the spectral sensing unit, and computational efficiency of the control algorithm—affect the system's response and distribution stability. Future research will focus on these issues to further refine the design and improve field adaptability of centrifugal variable-rate fertilizer spreaders under complex conditions.

References

1. Ban ZB, Xie H, He Y (2013) Experiment and simulation on response performance of electro-hydraulic VVT system for gasoline engine. Trans Chin Soc Agric Mach 44(9):12–18
2. Barros MMD, Volpato CE, Silva FMD et al (2016) Performance of a variable-rate distribution system for simultaneous fertilizer application. Rev Bras Eng Agric Ambient 20(2):188–192
3. Cai K, Xu X, Yu L et al (2016) Design of soil moisture sensor based on detection of propagation delay in LVDS differential transmission lines. Trans Chin Soc Agric Mach 47(12):315–322
4. Chattha HS, Zaman QU, Chang YK et al (2014) Variable rate spreader for real-time spot-application of granular fertilizer in wild blueberry. Comput Electron Agric 100(1):70–78
5. Chen M, Shi YY, Wang XC et al (2015) Expert decision system of variable nitrogen application in winter wheat. Trans Chin Soc Agric Mach 46(7):17–22
6. Liu YC (2012) Key technology on variable rate application system [D]. Chinese Academy of Agricultural Mechanization Sciences, Beijing
7. Su N, Xu TS, Song LT et al (2015) Variable rate fertilization system with adjustable active feed-roll length. Int J Agric Biol Eng 8(4):19–26
8. Su N, Xu T, Song L (2015) Development of a variable rate fertilization control system based on DC motor for use on granular fertilizer applicator [C]. In: 2015 Fifth International Conference on Communication Systems and Network Technologies (CSNT). IEEE, pp 1205–1209
9. Wang H, Bai XP, Liang HB (2017) Proportional distribution method for estimating actual grain flow under combine harvester dynamics. Int J Agric Biol Eng 10(4):158–164
10. Esquivel W, Cifuentes D, Ortega R (2015) Field testing of an automatic control system for variable rate fertilizer application. Comput Electron Agric 113:260–265
11. Lang CL, Wang JW, Wang JF et al (2013) Variable fertilizer control system for deep-fertilization liquid fertilizer applicator [J]. Trans Chin Soc Agric Mach 44(2):43–47
12. Liu KX, Zhang XQ, Bao WM et al (2015) A system response correction method with runoff error smooth matrix. J Hydraul Eng 46(8):960–966
13. Zhang CL, Wu R, Chen LQ (2017) Design and test of electronic control seeding system for maize. Trans Chin Soc Agric Mach 48(2):52–59
14. Maleki MR, Ramon H, De Baerdemaeker J et al (2008) A study on the time response of a soil sensor-based variable rate granular fertiliser applicator. Biosyst Eng 100(2):160–166
15. Meng ZJ, Zhao CJ, Fu WQ et al (2011) Prescription map identification and position lag calibration method for variable rate application of fertilizer. Trans Chin Soc Agric Mach 42(7):204–209
16. Qiu BJ, Li ZP, Wu H et al (2007) Experimental study on variable-rate spraying equipment response capability. Trans Chin Soc Agric Eng 23(11):148–152
17. Qiu BJ, Li K, Shen CJ et al (2010) Experiment on response characteristics of variable-rate continuous spraying system. Trans Chin Soc Agric Mach 41(9):32–36
18. Sun H, Chen Z M, Ge W J. Theory of machines and mechanisms[M]. Beijing: Higher Education Press, 2008
19. Shi YY, Chen M, Wang XC et al (2017) Analysis and experiment of fertilizing performance for precision fertilizer applicator in rice and wheat. Trans Chin Soc Agric Mach 48(7):97–103

20. Shi YY, Chen M, Wang XC et al (2015) Design and experiment of precision fertilizer applicator actuator of rice and wheat[J]. J South China Agric Univ 36(6):119–124
21. Wang XC, Chen M, Sun GX et al (2015) Design and test of control system on variable fertilizer applicator for winter wheat. Trans Chin Soc Agric Eng 31(S2):88–92
22. Yu XM, Jiang JP (2012) Adaptive networked control system based on delay prediction using neural network. J Zhejiang Univ (Eng Sci) 46(2):194–198
23. Zhang W (2013) Research on optimal design for multivariable systems based on internal model control theory [D]. Shanghai Jiao Tong University, Shanghai
24. Zhang YZ, Chen HT, Hou SY et al (2015) Design and experiment of slave computer control system for applying variable-rate liquid fertilizer. J Northeast Agric Univ (English Edition) 22(2):73–79
25. Zhang R, Zhao CJ, Wang X et al (2012) A variable-rate fertilizer control system for disc fertilizer spreaders. Intell Autom Soft Comput 18(5):461–467
26. Yang XJ, Zhang Y (2014) Modeling and analysis of the dynamics for the crank-slider mechanism considering the two pair clearances. Mech Sci Technol Aerosp Eng 33(11):1614–1620

Chapter 7
Research on Fertilizer Discharge Flow Rate Detection System

Yinyan Shi, Xiaochan Wang, J. Wang, S. Hu, Xuekai Huang, and C. Xia

Abstract To address the drawbacks of traditional granular fertilizer detection methods—such as high cost, low efficiency, and heavy labor demand—this study presents an intelligent detection system based on deep learning. The system integrates two core functions: automatic analysis of fertilizer deposition patterns and real-time flow rate detection in centrifugal spreaders. Its goal is to improve fertilization accuracy, optimize the structure of spreading devices and enable closed-loop control in variable-rate application. The hardware system includes a custom calibration mat, image acquisition unit, control terminal, and processing core for image capture, storage, and control. On the software side, two deep learning models are developed. The first is the F-TPP (Fertilizer-Swin Transformer PANet Partition) instance segmentation model, which combines Swin Transformer and PANet, and optimizes anchor scales to improve segmentation of small and densely distributed particles. The second is an improved lightweight YOLOv5s-seg model for real-time discharge monitoring. It uses HGNetV2 as the backbone, incorporates AKConv for size adaptability, and adds the iRMB attention mechanism to improve shape sensitivity. The detection head is also refined to reduce parameters and expand the receptive field. Experimental results show that the F-TPP model achieves a precision of 93.8%, recall of 96.3%, and F1 score of 0.95, outperforming models like Mask R-CNN, SOLO, and YOLOv5. The improved YOLOv5s-seg model achieves 45 FPS, 95.9% mAP, a model size of 6.66 MB, and a computational cost of 17.8 G. Under fertilizer outlet openings of 17–25 mm and a flow range of

Y. Shi (✉) · X. Wang
College of Engineering, Nanjing Agricultural University, Nanjing, China

Jiangsu Province Engineering Lab for Modern Facility Agriculture Technology & Equipment, Nanjing, China
e-mail: shiyinyan@njau.edu.cn; wangxiaochan@njau.edu.cn

J. Wang · S. Hu · X. Huang · C. Xia
College of Engineering, Nanjing Agricultural University, Nanjing, China
e-mail: wangjihao@stu.njau.edu.cn; 2023112024@stu.njau.edu.cn; 2022212011@stu.njau.edu.cn; 9213010210@stu.njau.edu.cn

Y. Shi et al. (eds.), *Technological Progress on Variable Rate Fertilization*, Smart Agriculture 17, https://doi.org/10.1007/978-981-95-4330-4_7

15.13–54.83 g/s, the system reached a correlation coefficient of 0.9984 between measured and actual mass, with a maximum error of 5.07%.In conclusion, the proposed system provides a lightweight, accurate, and real-time solution for granular fertilizer monitoring, supporting intelligent variable-rate fertilization in precision agriculture.

Keywords Granular fertilizer detection · Variable-rate fertilization · Instance segmentation · YOLOv5s-seg · Deep learning · Flow rate detection

7.1 Introduction

The effective utilization of agricultural fertilizers is critical for enhancing agricultural productivity, and the proper application of granular fertilizers is essential to maximize this effectiveness. However, in many rural regions of China, farmers still rely on experience-based manual fertilization, and excessive, unregulated fertilization remains a prevalent issue. This leads to an imbalanced fertilizer composition and low utilization efficiency, which in turn contribute to environmental problems such as soil and water pollution. With the government's promotion of precision agriculture, various types of agricultural machinery and equipment are replacing manual labor in farming operations [1, 2]. Precision agriculture has been widely adopted in crop management—particularly in practices like accurate fertilizer application and seed sowing—delivering substantial benefits. Granular compound fertilizers, which are rich in essential elements (e.g., nitrogen, phosphorus, and potassium) and other nutrients crucial for crop growth, possess unique physical properties: they can flow like a fluid in fertilizer tanks while retaining their solid form. These characteristics make granular fertilizers well-suited for mechanized application. As a core technology in precision agriculture, variable-rate fertilization enables fertilizer applicators to precisely regulate the amount of fertilizer applied based on pre-set or real-time generated prescription maps. This technique improves fertilizer utilization and reduces spatial variation in soil nutrient distribution, which is vital for boosting agricultural economic efficiency and promoting sustainable, healthy agricultural development [3].

Highly accurate and dynamic monitoring of solid granular fertilizer flow is indispensable for closed-loop control in variable-rate fertilization systems and for real-time evaluation of operational performance. To address the challenge of real-time detection of solid granular fertilizer flow, researchers at home and abroad have carried out extensive studies. Existing technologies for granular fertilizer flow detection—used both in practical applications and research—include fertilizer tank weighing, indirect measurement of fertilizer discharge shaft speed, electrostatic induction, photoelectric detection principles, microwave technology, and image analysis. One approach involved designing a dynamic weighing system for on-field fertilizer applicators, which operated by measuring the difference between the initial mass of fertilizer and the remaining amount. The fertilizer discharge flow rate was estimated by establishing a model linking the rotational speed of the discharge shaft to a pre-calibrated fertilizer flow model [4]. However, this method was susceptible to performance variations between different fertilizers and machine malfunctions, which impaired its detection accuracy. Another solution introduced a new sensor based on capacitance technology for online detection of fertilizer quantities. Researchers modeled the relationship between the mass flow rates of nitrogen (N), phosphorus (P), and potassium (K) fertilizers and the capacitance output to minimize the impact of flow pulsation on fertilizer application rates [5]. A third method proposed a granular fertilizer flow detection system based on the principle of light quantity blocking. This approach effectively addressed the issue of capturing flow information, especially under high-flow conditions where particles block each other during fertilizer application. Additionally, the Doppler effect principle was integrated into a mathematical model that correlated Doppler echo signals with particle falling velocity and fertilizer concentration in the pipeline. The application of Kalman filtering further improved the precision of flow rate measurements for solid fertilizer particles [6–8].

With advancements in computer science, image processing technology has enabled real-time, clear visualization of the spatial distribution and dynamic flow processes of particulate fertilizers. Researchers developed a high-resolution, high-speed particle projection imaging system that accurately analyzes the spatial distribution of granular fertilizers, automatically calculates the velocity and direction of each particle, and estimates their trajectories. Another system was designed to measure fertilizer flow using image processing techniques, which accurately separated granular fertilizer from the background in a two-dimensional projection plane [9, 10]. This system controlled fertilizer application rates using an open-loop controller and a PID controller, and the fertilizer flow rate estimated by the imaging system showed good agreement with the value obtained through actual weighing methods. Experimental and image processing techniques were also used to investigate the flow rate at the bottom outlet of a vertical two-dimensional silo, revealing the relationship between flow rate and time by counting instantaneous flow data. Furthermore, an image-based mass flow measurement system for plastic pellets was developed; this system captured images of pellets on a conveyor belt at regular intervals and achieved high-precision detection of mass flow rates. To address challenges in accurately segmenting small granular fertilizers and densely arranged particles (thereby improving the accuracy of automatic fertilization), a deep learning segmentation model for granular fertilizers—named Fertilizer-Swin-Transformer-PANet—was proposed [11].

While existing research on fertilizer discharge flow detection primarily focuses on outer groove wheel-type band fertilizer applicators or low-flow-rate fertilizer dischargers, centrifugal fertilization in high-flow, high-efficiency systems poses unique challenges. In such systems, when particles fall from the discharge port to the spreading disks, they tend to overlap, making it difficult for traditional image processing algorithms to detect individual particles. Moreover, current flow measurement systems often lack the speed and accuracy required to handle fast-moving, dense granular fertilizer flows. This underscores the need for more advanced solutions to accurately measure discharge flow rates in centrifugal fertilizer spreaders. Compared with traditional physical sensors, image processing technology enables non-contact particle measurement, reduces wear and tear, and eliminates reliance on specific signal processing techniques—making it more adaptable to diverse operating scenarios. It also enhances system reliability and provides high-resolution, real-time visualization of fertilizer particle distribution and flow dynamics, offering significant advantages over conventional methods. Additionally, when integrated with advanced deep learning models, image processing technology can efficiently handle the complexity of high-density, fast-moving particle flows, enabling accurate detection of small or overlapping particles [12, 13].

The bench test setup for measuring the fertilizer discharge flow rate of the centrifugal variable-rate applicator consisted of a fertilizer tank, an image acquisition device, a discharge port control unit, a PC (Lenovo Legion R9000P), and a fertilizer collection and weighing system, as illustrated in Fig. 7.1. The image acquisition device comprised LED spotlights, a high-speed black-and-white industrial camera (U3-33GC), and an inspection enclosure, which together ensured high-quality images of granular fertilizers. The discharge port control unit employed a stepping motor (35CM04) to adjust the outlet opening, thereby regulating the flow rate. Image frames captured during operation were processed on the PC, where the image processing module segmented the particles, estimated their total mass, and calculated the fertilizer flow rate corresponding to each discharge port setting. The collection and weighing system, consisting of a fertilizer box, buffer pad, and electronic

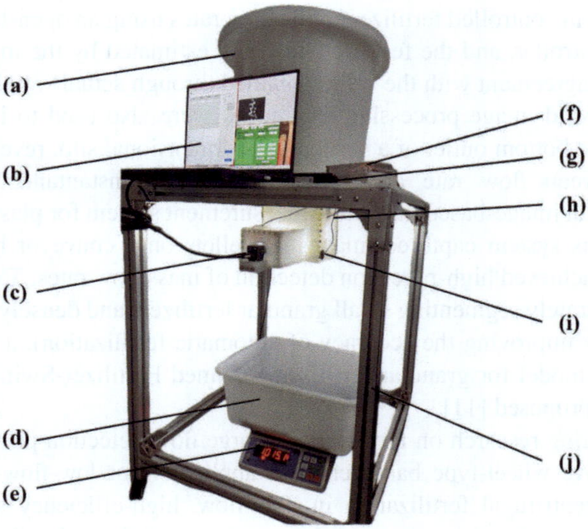

Fig. 7.1 Experimental platform for the centrifugal variable fertilizer discharge flow detection system: (**a**) Fertilizer box. (**b**) PC. (**c**) Camera. (**d**) Fertilizer collection device. (**e**) Weighing device. (**f**) Power supply. (**g**) Drive controller. (**h**) Fertilizer discharge port control device. (**i**) LED spotlights. (**j**) Test bench

scale (AnHen-ZCS), was used to weigh the discharged fertilizer, providing reference measurements for actual flow rate determination.

7.2 Material and Methods

The bench test setup for measuring the fertilizer discharge flow rate of the centrifugal variable-rate applicator consisted of a fertilizer tank, an image acquisition device, a discharge port control unit, a PC (Lenovo Legion R9000P), and a fertilizer collection and weighing system, as shown in Fig. 7.1. The image acquisition device included LED spotlights, a high-speed black-and-white industrial camera (U3-33GC), and an inspection enclosure, which together ensured the capture of clear images of granular fertilizers. The discharge port control unit employed a stepping motor (35CM04) to adjust the opening of the fertilizer outlet, thereby regulating the flow rate. Image frames acquired during operation were processed on the PC, where the image processing module segmented the particles, estimated their total mass, and calculated the fertilizer flow rate corresponding to each discharge port setting. The collection and weighing system, comprising a fertilizer box, buffer pad, and electronic scale (AnHen-ZCS), was used to weigh the discharged fertilizer, providing the reference values for actual flow rate measurement.

7.2.1 *Basic Parameters of Granular Fertilizers*

Granular compound fertilizers offer advantages such as approximately spherical shape, convenient storage, balanced nutrient content, strong mobility, and effective spreading—making them widely used in centrifugal variable-rate fertilizer discharge devices. In this study, white granular compound fertilizers containing nitrogen, phosphorus, and potassium were selected as experimental materials; these fertilizers are commonly applied in the rice-wheat rotation areas of Jiangsu Province. A digital vernier caliper (accuracy: 0.01 mm) was used to measure the triaxial dimensions of the fertilizer granules, namely length (L), width (B), and thickness (H). The shape of the granular fertilizer is shown in Fig. 7.2a. The equivalent diameter was calculated using the method described by $D = \left(L \times B \times H\right)^{1/3}$ [6], while the sphericity of the fertilizer was computed using the method described by $S = D/L \times 100\%$. The frequency distribution of equivalent diameters for a sample of 100 particles was analyzed, with results presented in Fig. 7.2b. Statistical analysis revealed that the equivalent diameters followed a normal distribution, with an average value of 3.88 mm, a maximum of 4.72 mm, and a minimum of 2.99 mm. The sphericity of the fertilizers exceeded 90%, allowing them to be approximated as spherical in experimental calculations. The density of the fertilizer particles was measured via the liquid displacement method, yielding a value of 1.216 g/cm^3.

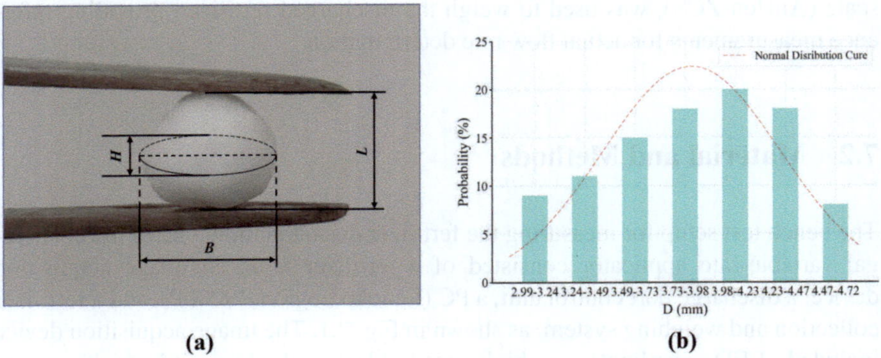

Fig. 7.2 Shape and equivalent diameter distribution of fertilizer: (**a**) Granular fertilizer of shape. (**b**) Equivalent diameter distribution of granular fertilizer

Additionally, the average moisture content, angle of repose, average crushing force, and static friction coefficient of the compound fertilizers were 1.02%, 23.5°, 4.67 N, and 0.486, respectively. To minimize the impact of environmental conditions on the fluidity and hygroscopicity of the fertilizer granules, all experiments were conducted under temperatures ranging from 18–25 °C and a relative humidity of 35–50%.

7.2.2 *Image Acquisition System*

7.2.2.1 Acquisition of Particle Images

The image acquisition device was specifically designed to handle the rapid descent and dense distribution of granular fertilizers. The falling fertilizer stream was enclosed in a cassette with a dark internal environment to reduce external light interference. LED spotlights provided consistent, uniform illumination to ensure stable lighting conditions. The high-speed camera, which can capture 1000 frames per second, was sufficiently fast to track the movement of the rapidly falling particles. This setup ensured that the fertilizer particles remained well-lit at all times, enabling the capture of clear, high-contrast images even during the particles' rapid descent.

7.2.2.2 Calculation of Particle Transit Time Through the Camera's Field of View

To initially determine the time required for particles to pass through the camera's field of view, the camera-based collection process for granular fertilizers is illustrated in Fig. 7.3a. When particles fall from the fertilizer tank, pass through the detection area, and move toward the fertilizer spreading disc, they are subjected to

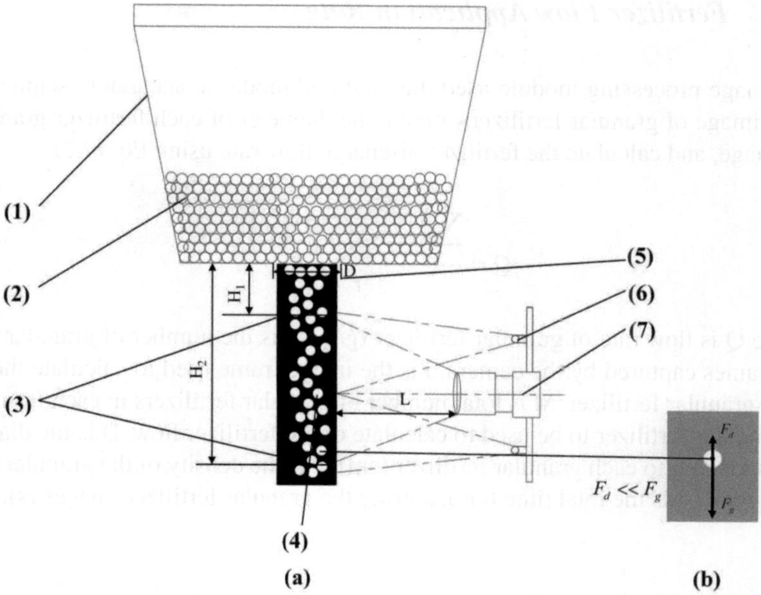

Fig. 7.3 The motion of granular fertilizer as it falls from the fertilizer tank and passes through the camera's field of view: (**a**) Camera-based collection of granular fertilizers: (1) Fertilizer bin; (2) Granular fertilizer; (3) Camera acquisition field of view; (4) Targeted granular fertilizer stream; (5) Fertilizer discharge opening; (6) LED spotlights; (7) Camera. (**b**) Force analysis of a single granular fertilizer particle

the forces of gravity (F_g) and resistance (F_d), as shown in Fig. 7.3b. During the particles' motion, their speed continuously increases due to gravity, and force F_d also increases with velocity. However, the magnitude of force F_d is far smaller than that of force F_g acting on the particles, making its impact on particle motion negligible. Therefore, for the sake of simplicity, force F_d was ignored in this process. It was assumed that the granular fertilizer undergoes free-fall motion from the fertilizer tank's discharge port to the fertilizer spreading disc. Equation (7.1) was used to calculate the time (T) between the appearance and disappearance of granular fertilizer in the camera's field of view.

$$\mathrm{T} = \sqrt{\frac{2}{g}}\left(\sqrt{H_2} - \sqrt{H_1}\right) \tag{7.1}$$

In the equation, H_2 denotes the distance from the bottom of the camera's field of view to the fertilizer tank's discharge port, while another H_1 represents the distance from the top of the camera's field of view to the discharge port; g denotes the gravitational acceleration (0.0098 mm/ms^2). In this study, values of H_2 and H_1 were 137 mm and 86 mm, respectively, leading to a preliminary calculation of T as 35 ms.

7.2.3 Fertilizer Flow Application Rate

The image processing module used the modified model to accurately segment the mask image of granular fertilizers, obtain the diameter of each fertilizer granule in the image, and calculate the fertilizer discharge flow rate using Eq. (7.2).

$$Q = \frac{\sum_{i=1}^{N}\sum_{j=1}^{M}\left(\frac{\pi}{6}D_{ji}^{3}\rho\right)}{t} \tag{7.2}$$

Where Q is flow rate of granular fertilizer (g/s). N is the number of granular fertilizer frames captured by the camera. i is the image frame used to calculate the flow of the granular fertilizer. M is total number of granular fertilizers in each frame. j is the granular fertilizer to be used to calculate of the fertilizer flow. D is the diameter corresponding to each granular fertilizer (cm). ρ is the density of the granular fertilizer (g/cm^3). t is the total time for acquiring the granular fertilizer images (s).

7.2.4 Data Set Construction

At different discharge port openings, the granular fertilizers captured by the camera appeared in three distinct states: non-overlapping, partially overlapping, and densely overlapping, as illustrated in Fig. 7.4. A total of 1000 images were acquired at a resolution of 496 × 308 pixels, which was sufficient to ensure clear fertilizer particle capture. The images were manually annotated using Labelme, and to improve dataset diversity and labeling robustness, data augmentation techniques—including rotation, brightness adjustment, and Gaussian blurring—were applied. This process expanded the dataset to 4500 images, which were subsequently divided into training, validation, and test sets in an 8:1:1 ratio.

7.2.5 YOLOv5s Mold

YOLOv5s is a widely used single-target detection algorithm, noted for its fast training, lightweight design, and ease of deployment on mobile platforms [17]. As illustrated in Fig. 7.5, the network comprises four main modules: Input, Backbone, Neck, and Detect. The Input module employs the Mosaic augmentation technique [18], which randomly scales, crops, and rearranges images to enhance recognition performance. The Backbone incorporates Focus and Conv structures to downsample high-resolution fertilizer images into multi-level, low-resolution feature maps. The Neck leverages a Feature Pyramid Network (FPN) [25] to pass high-level semantic information through top-down upsampling [27]. Finally, the Detect

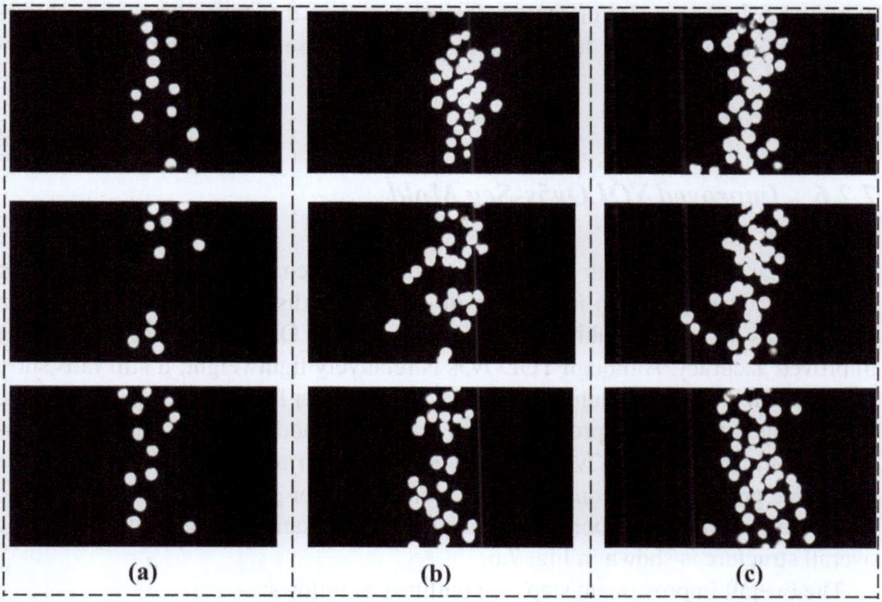

Fig. 7.4 The three states between granular fertilizers in the image: (**a**) No overlap. (**b**) Partial overlap. (**c**) Dense overlap

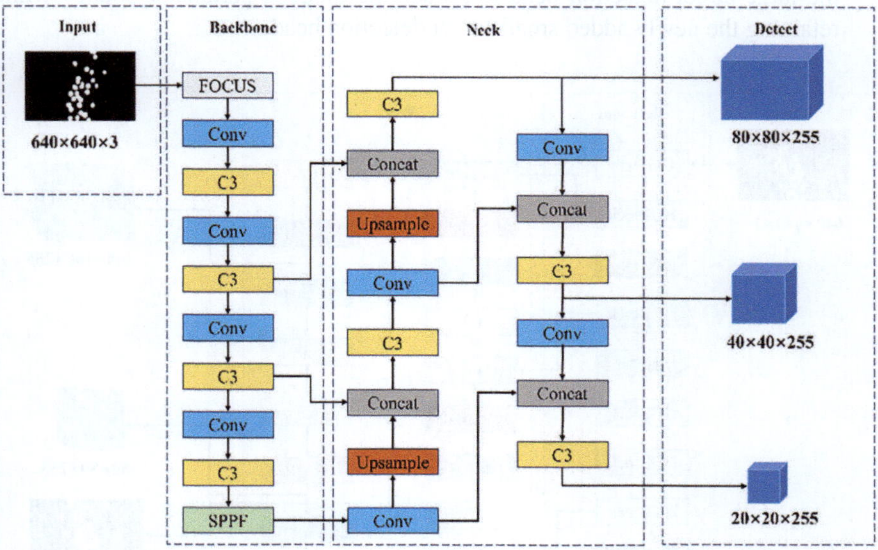

Fig. 7.5 YOLOv5s network architecture

module applies non-maximum suppression to select the optimal bounding boxes corresponding to granular fertilizers, outputting precise location and category predictions [4, 19].

7.2.6 Improved YOLOv5s-Seg Mold

As granular fertilizer exits the outlet, particle flow characteristics cause varying degrees of overlap, which intensify with increasing discharge rates. This overlap complicates detection, making enhancements to the YOLOv5s model essential for improved accuracy. Although YOLOv5s is relatively lightweight, it still falls short of mobile deployment requirements, particularly under high flow conditions where rapid image capture and processing are critical. To address these challenges, this study developed the YOLOv5s-seg network, derived from YOLOv5. The improved model integrates an additional segmentation branch for granular fertilizer semantic segmentation while incorporating lightweight optimization strategies [15, 16]. The overall structure is shown in Fig. 7.6.

The overall improvement steps are outlined as follows:

1. Replace the original backbone network with the more lightweight HGNetV2 structure.
2. Increase the number of original detection heads from three to four, then remove the large-target detection head to reduce the model's parameter count—while retaining the newly added small-target detection head.

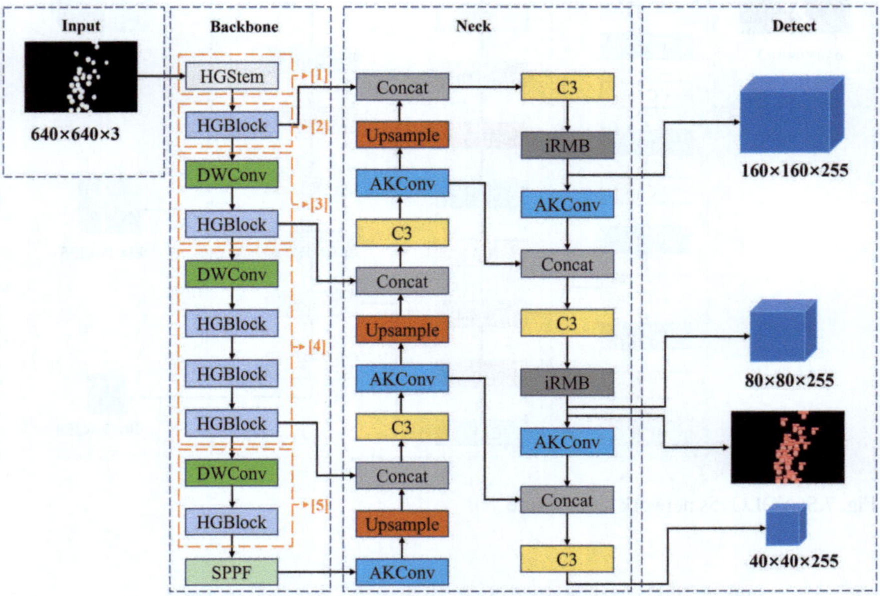

Fig. 7.6 Improved YOLOV5s-seg network architecture

Table 7.1 Analysis of typical lightweight network performance

Lightweight network	Size of model/MB	Computaion/G
ShuffleNetV2	2.3	146
SqueezeNet-1.0	1.25	833
GhostNet	5.4	141
EfficientNet-B0	5.31	390
MobileNetV3-large	5.38	219
HGNetV2	2.64	52

3. Adopt the AKConv (Arbitrary Kernel Convolution) structure.
4. Incorporate the iRMB attention mechanism.

7.2.6.1 Lightweight Backbone Network Improvements

To optimize the backbone network of YOLOv5s-seg, this study compared the performance of various mainstream lightweight networks currently available (Table 7.1). Among these networks, HGNetV2 stood out due to its minimal computational demands while maintaining strong performance, achieving an effective balance between memory constraints and detection speed. Consequently, HGNetV2 was selected to replace the original backbone of YOLOv5s-seg.

The HGNetV2 architecture [28] is tailored for GPU platforms, adopting a design strategy that prioritizes 3 × 3 standard convolutions to achieve high computational efficiency. As illustrated in Fig. 7.7a, its initial Stem module acts as the preprocessing stage, extracting primary features from raw fertilizer particle inputs. With 96 channels, this layer reduces both parameter count and computational cost. The network is composed of four sequential HG Stages, each built from HG Blocks that rely heavily on standard convolution operations. Figure 7.7b shows that these blocks hierarchically process information, enabling the model to capture both shallow and deep feature representations of fertilizer granules. Layers three through five introduce learnable down-sampling (LDS) modules, which shrink the spatial size of feature maps to lower computational demand while broadening the receptive field. In this work, YOLOv5s-seg's backbone was replaced by HGNetV2 (Fig. 7.7c). The revised structure begins with the HG Stem, followed by four HG Stages, with LDS layers in the third to fifth stages enhanced by Depthwise Convolution (DWConv). The final feature map is then fed into the SPPF (Spatial Pyramid Pooling—Fast) module. This replacement reduces memory and computation overhead, improving efficiency and making the network more deployable on mobile devices.

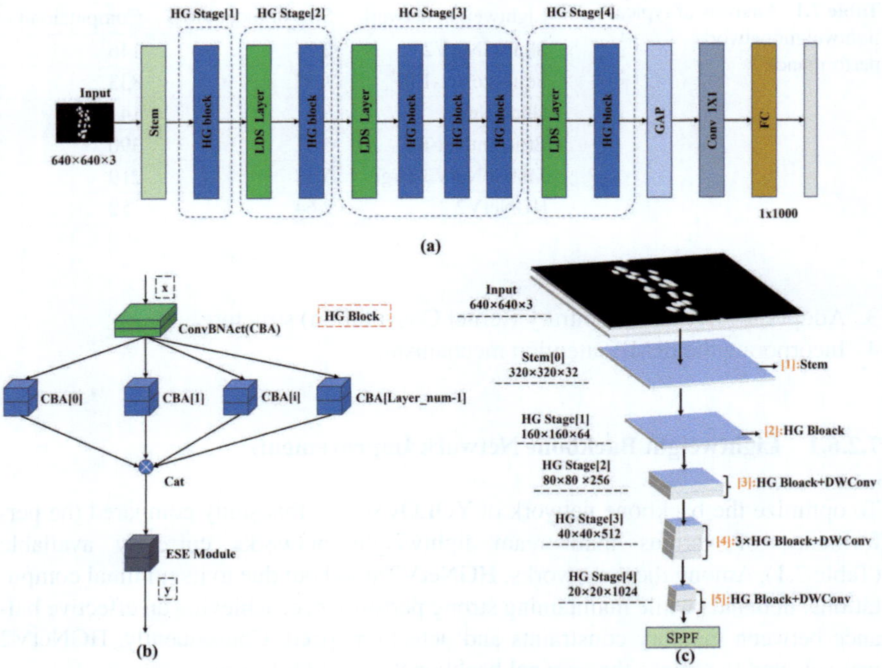

Fig. 7.7 HGNetV2 network structure diagram: (**a**) The HGNetV2 framework. (**b**) The HG Block structure. (**c**) The replaced YOLOV5s-seg backbone network

7.2.6.2 Detection Head

Increasing the number of detection heads in the YOLOv5s-seg model from three to four [14] expands the receptive field, allowing the model to capture more detailed low-level feature information of granular fertilizers. This modification enhances the model's detection accuracy, particularly for small granular fertilizers. Additionally, to reduce the overall model size, the large-target detection head was removed—while the newly added small-target detection head was retained—to ensure better performance in detecting smaller particles.

7.2.6.3 The AKConv Structure

The Conv convolution layer in the Neck structure was replaced with the lightweight AKConv structure (Fig. 7.8). AKConv assigns initial sampling coordinates to a convolution of arbitrary size and adjusts the sampling shape using learnable offsets. Unlike fixed sampling shapes, the sampling shape at each position is modified through resampling. As shown in Fig. 7.9a, the convolution kernel size in AKConv is a flexible parameter, which can be set to any desired value (e.g., 2, 3, 4, 5, 6, etc.). Beyond the flexibility in parameter count selection, the initial shape of the

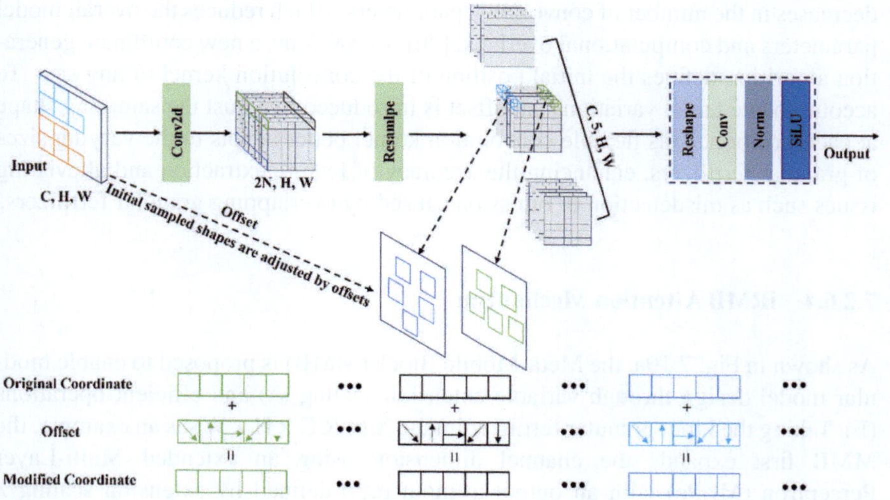

Fig. 7.8 The detailed schematic of the structure of AKConv

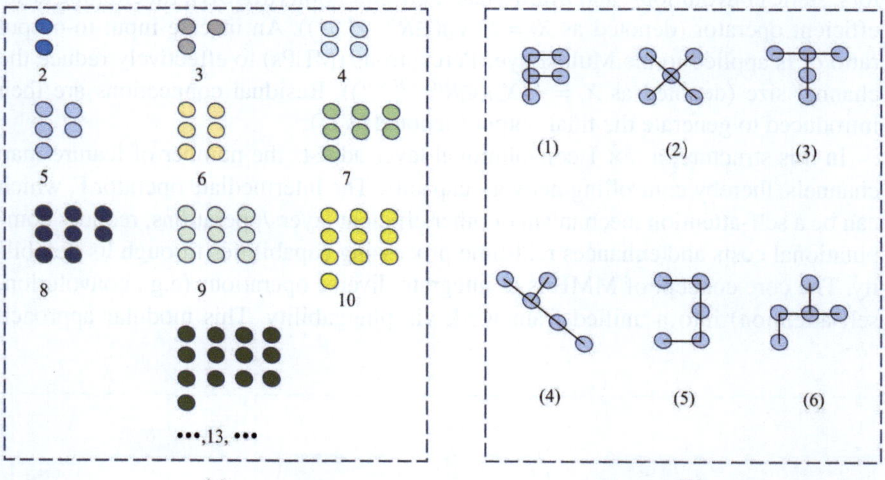

Fig. 7.9 The AKConv with arbitrary convolutional kernel sizes and shapes: (**a**) The initial sampled coordinates for arbitrary convolutional kernel sizes. (**b**) The arbitrary sampling shapes of different initial sample shape of size 5

convolution kernel is also customizable; Fig. 7.9b presents the initial design of a convolution kernel with 5 parameters. This structure provides a flexible convolution mechanism, enabling the kernel to adjust both parameter count and sampling shape. By modifying the kernel's size and shape, the model can dynamically adapt its receptive field to capture different features of large and small particles, thereby improving detection accuracy. The AKConv structure supports linear increases or decreases in the number of convolution parameters, which reduces the overall model parameters and computational overhead [26]. In AKConv, a new coordinate generation algorithm defines the initial position of the convolution kernel of any size. To accommodate target variations, an offset is introduced to adjust the sampling shape at each position. This flexible convolution kernel better adapts to the varying sizes of granular fertilizers, enhancing the accuracy of feature extraction and alleviating issues such as misdetection or omission caused by overlapping granular fertilizers.

7.2.6.4 iRMB Attention Mechanism

As shown in Fig. 7.10a, the Meta-Mobile Block (MMB) is proposed to enable modular model design through variable extension scaling (λ) and efficient operations (F). Taking the input granular fertilizer image X (∈R C × H × W) as an example, the MMB first expands the channel dimension using an extended Multi-Layer Perceptron (MLPe) with an output-to-input ratio defined by extension scaling λ (denoted as $X_e = MLPe(X)(\in R^{\lambda C \times H \times W})$). The intermediate operator F enhances image features by integrating various efficient operations, including constant operators, static convolutions, and Multi-Head Self-Attention (MHSA), functioning as an efficient operator (denoted as $X_f = F(X_e)(\in R^{\lambda C \times H \times W})$). An inverse input-to-output ratio of is applied to the Multi-Layer Perceptrons (MLPs) to effectively reduce the channel size (denoted as $X_f = F(X_e)(\in R^{\lambda C \times H \times W})$). Residual connections are then introduced to generate the final output (denoted as D).

In this structure, a 1 × 1 convolutional layer adjusts the number of feature map channels, thereby controlling network capacity. The intermediate operator F, which can be a self-attention mechanism or other efficient layers/operations, reduces computational costs and enhances real-time processing capabilities through its flexibility. The core concept of MMB is to integrate diverse operations (e.g., convolution, self-attention) into a unified framework via pluggability. This modular approach

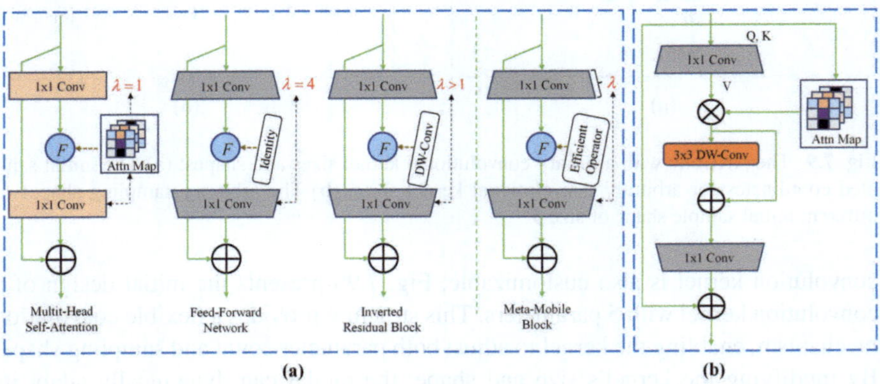

Fig. 7.10 Structural diagram of the inverted residual block with attention: (**a**) The Meta-Mobile Block. (**b**) The iRMB

allows the model to adjust its capacity on demand without redesigning the entire network, improving efficiency and flexibility.

Building on MMB, an inverse residual mobile block (iRMB) is developed, combining the lightweight properties of convolutional neural networks with the dynamic processing capabilities of Transformer models for intensive prediction tasks on mobile devices [29]. The iRMB structure is illustrated in Fig. 7.10b, where F is modeled as a cascade of MHSA and DWConv layers (denoted as $F(\cdot) = (DWConv, Skip)(MHSA(\cdot)))$. It extends to an attention-based model using inverted residual blocks, enhancing the model's ability to extract key features when processing visual and sequential data. This design enables the model to capture edge details in dense, overlapping regions of fertilizer particles by selectively focusing on relevant input components while considering the entire input space. The modular design allows rapid adaptation and optimization for different environments and tasks, achieving a flexible balance between capacity and efficiency. Consequently, iRMB ensures high computational performance for demanding applications like real-time granular fertilizer flow detection, with improved efficiency and precision.

7.3 Experiment and Analysis

7.3.1 Test Environment

The experiment was conducted on a Windows 11 operating system with hardware configuration: AMD Ryzen 7-5800H processor and NVIDIA GeForce RTX 3060 Laptop GPU. The programming environment used Python 3.8 with the PyTorch 1.8.1 deep learning framework. Training parameters were set as follows: initial learning rate = 0.0001 (adjusted via cosine-annealing schedule with an attenuation coefficient of 0.0005) and iteration count = 1000.

7.3.2 Indicators for Model Performance Assessment

In this work, the evaluation of granular fertilizer segmentation performance was based on Precision (P), Recall (R), mean average precision (mAP), and Total-loss [5, 30]. To comprehensively assess efficiency, the number of network parameters and the detection-segmentation rate, expressed in frames per second (FPS) for single-image inference, were also considered [14]. The definitions of FPS, mAP, and Average Precision (AP) are shown in Eqs. (7.3)–(7.5). As illustrated in Fig. 7.11, the modified YOLOv5s-seg achieved smoother training curves with markedly reduced gradient fluctuations [9, 10]. Moreover, the optimized model demonstrated faster convergence, higher Preci

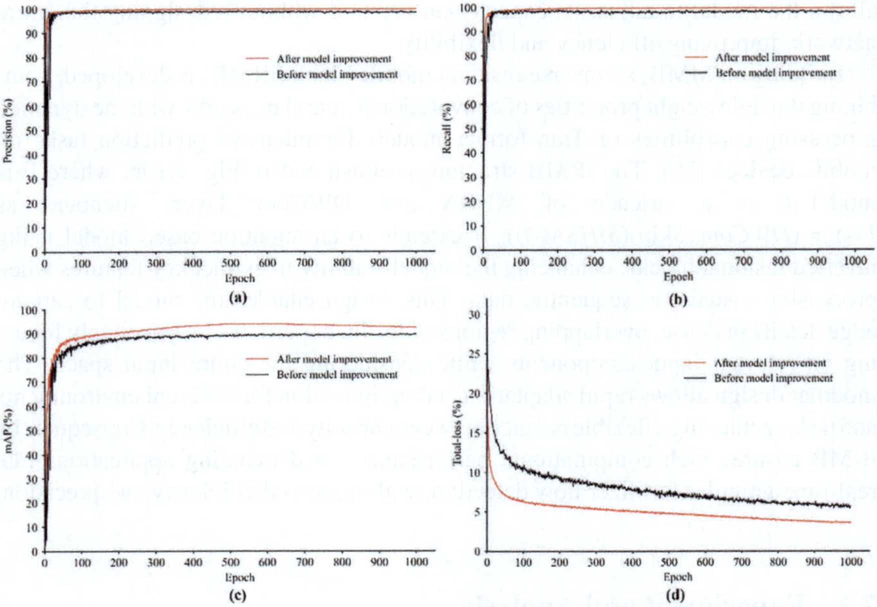

Fig. 7.11 The curve of training process before and after improvement of YOLOv5s-seg segmentation training procedure: (**a**) Precision curve. (**b**) Recall curve. (**c**) mAP curve. (**d**) Total-loss curve

$$\mathrm{FPS} = \frac{1000}{P_{reTime} + I_{nferTime} + N_{MSTime}} \tag{7.3}$$

$$\mathrm{mAP} = \frac{\sum_{x=0}^{n} \int_0^1 P(R)\,dR}{n} \tag{7.4}$$

$$\mathrm{AP} = \int_0^1 P(R)\,dR \tag{7.5}$$

Where P_{reTime} = image preprocessing time (ms), $I_{nferTime}$ = network inference time (ms), N_{MSTime} = network prediction frame optimization time (ms), and n = number of detected categories.

The total loss function of the YOLOv5s network includes classification loss (L_{cls}), localization loss (L_{box}), and confidence loss (L_{obj}). Equation (7.6) represents the total loss function of YOLOv5s-seg (L_{total}), incorporating an additional semantic segmentation loss (L_{seg}). Equation (7.7) defines localization, classification, and segmentation losses using binary cross-entropy, while Eq. (7.8) represents confidence loss.

$$L_{total} = L_{cls} + L_{box} + L_{obj} + L_{seg} \tag{7.6}$$

$$L_{box,cls,seg} = -\frac{1}{TN}\sum_{a}\left[b\ln z + (1-b)\ln(1-c)\right] \tag{7.7}$$

$$L_{obj} = 1 - \frac{\left|B \cap B^{gt}\right|}{\left|B \cup B^{gt}\right|} + \frac{\left|C - B \cup B^{gt}\right|}{C} \tag{7.8}$$

Where a is the sample, b is the label score, c is the predicted output score, and TN is the total number of samples. B^{gt} denotes the true regression frame, B denotes the predicted regression frame, and C is the case of overriding B and B^{gt} no overlap, in which the predicted frame eventually regresses to the target frame.

7.3.3 Model Performance Analysis

7.3.3.1 Analysis of Ablation Test Results

In this experiment, the lightweight HGNetV2 backbone significantly reduced model parameters. The 新增 detection head expanded the network's receptive field, and the AKConv structure enabled adaptation to varying sizes and shapes of granular fertilizers. Additionally, an iRMB attention mechanism was integrated before the detection head and segmentation network branches. Eight experimental groups were evaluated using mAP, model size, and FPS as key metrics.

Table 7.2 presents ablation experiment results: 1) Replacing the backbone with HGNetV2 improved mAP by 1.7% and reduced model size by 4.75 MB. 2) Modifying the detection head alone enhanced small-target detection accuracy and increased FPS by 2 frames/s. 3) Adjusting the AKConv structure (enabling adaptive convolution kernels for fertilizer particle shapes/sizes) increased mAP by 3.4%. 4) Adding the iRMB attention mechanism improved mAP by 6.2% and increased FPS by 5 frames/s.

Table 7.2 Results of ablation experiments using the improved YOLOv5s-seg

HGNetV2	Detector head	AKConv	iRMB	mAP/%	Size of model/MB	FPS/(frames/s)
✗	✗	✗	✗	89.7	14.4	40
✓	✗	✗	✗	91.4	9.65	48
✗	✓	✗	✗	92.5	11.5	42
✗	✗	✓	✗	93.1	11.6	41
✗	✗	✗	✓	94.8	14.1	37
✓	✓	✗	✗	93.5	7.15	43
✓	✓	✓	✗	94.9	6.95	47
✓	✓	✓	✓	95.9	6.64	45

Note: "✗" is not added to this module; "✓" is added to this module

Table 7.3 Performance comparison between different YOLOv5 versions and the improved YOLOv5s-seg

Model	AP/%	mAP/%	Computaion/G	Size of model/ MB	FPS/ (frames/s)	*F*1
YOLOv5n-seg	95.3	88.4	6.4	7.74	60	0.946
YOLOv5s-seg	97.9	89.7	25.9	14.41	40	0.977
YOLOv5m-seg	98.2	91.1	70.2	31.52	28	0.981
YOLOv5l-seg	98.4	91.9	147.1	78.75	21	0.985
YOLOv5x-seg	98.6	93.4	264.9	168.91	17	0.989
Improved YOLOv5s-seg	99.9	95.9	17.8	6.64	45	0.998

These modifications enhanced sensitivity to overlapping targets, accelerated convergence, and stabilized detection performance. The final improved model achieved optimal granular fertilizer segmentation results with significant gains in accuracy, speed, and stability.

7.3.3.2 Performance Comparison of Improved YOLOv5s-Seg with Other YOLOv5 Series

YOLOv5-seg comprises five variants: YOLOv5n-seg, YOLOv5s-seg, YOLOv5m-seg, YOLOv5l-seg, and YOLOv5x-seg. As summarized in Table 7.3, the enhanced YOLOv5s-seg achieved an AP of 99.9%, an mAP of 95.9%, an F1 score of 0.998, a computational load of 17.8 G, a model size of 6.64 MB, and an inference speed of 45 FPS. Compared with YOLOv5n-seg, it improved AP, mAP, and F1 by 4.6%, 7.5%, and 0.052, respectively, while reducing size by 1.1 MB. Against the larger models (YOLOv5m/l/x-seg), it reduced model size by 78.9–96.1% and computational demand by 70.5–93.1%, while raising FPS by 17–28 frames/s. Relative to the baseline YOLOv5s-seg, the improved model increased AP and mAP by 2.0% and 6.2%, boosted FPS by 5, and simultaneously cut size and computation by 53.9% and 67.6%. Overall, the optimized YOLOv5s-seg surpassed all other variants in balancing detection accuracy, computational efficiency, and lightweight deployment—demonstrating strong suitability for granular fertilizer image segmentation on mobile platforms.

7.3.3.3 Performance Comparison of Granular Image Segmentation Models

Figure 7.12 displays segmentation results of granular fertilizers obtained at three discharge port openings (representing varying overlap levels) using the original and improved YOLOv5s-seg models, as well as other YOLOv5-based segmentation networks. The outputs of the unimproved models exhibited problems such as

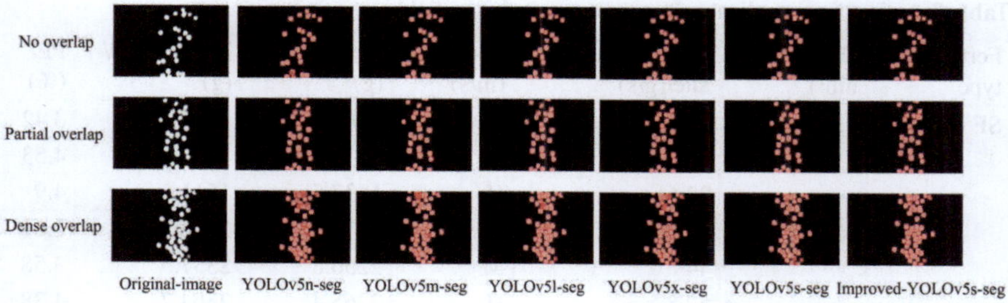

Fig. 7.12 Fertilizer segmentation effect diagram of the improved YOLOv5s-seg with

undetected small particles, false detection in overlapping regions, and imprecise particle mask boundaries. By contrast, the optimized YOLOv5s-seg network effectively recognized all fertilizer particles in the images, mitigating the errors caused by contour occlusion in overlapping cases. Even under conditions of severe overlap, it accurately determined particle counts and generated masks with edge contours closely aligned to the true particle shapes.

7.3.4 Field Evaluation of a Detection System for Discharge Flow Rate in a Centrifugal Variable Fertilizer Spreader

The detection device adopts a compact, integrated structure with dimensions of 280 mm × 240 mm × 200 mm. Owing to its lightweight design, it can be conveniently installed on the fertilizer spreader without affecting normal operation. In the 2FV-1500 centrifugal variable-rate fertilizer spreader, fertilizer granules drop from the tank and pass through the detection unit under illumination. A camera acquires images of the particles and transmits them to the Jetson Nano, where an improved algorithm processes the data. The terminal displays the fertilizer flow rate in real time and regulates discharge by adjusting the opening of the fertilizer outlet.

The Ward Auron 0X1604 tractor, fitted with a centrifugal variable-rate fertilizer spreader, was preset through the terminal to operate at three speeds: high (H, 1.2 m/s), medium (M, 1.0 m/s), and low (L, 0.8 m/s). It maintained a constant velocity over a 50 m test distance. The target fertilizer application rate, obtained from prescription maps with a 16 m spreading width, was combined with Eq. (7.9) to determine the required discharge flow rate, which governed the adjustment of the fertilizer outlet opening [20]. For application rates of 35 kg/hm^2 and 55 kg/hm^2, fertilizer flow was monitored using a single-camera small-flow method (SF) [21, 22], while rates of 140 kg/hm^2 and 200 kg/hm^2 were measured with a dual-camera large-flow method (DF). As presented in Table 7.4, the maximum detection error was less than 4.9% for SF and 5.04% for DF. Under high-speed conditions, relative error increased compared with low-speed operation, likely due to stronger machine

Table 7.4 Fertilizer application rate test results in the field

Fertilizer type	Target/(kg/hm^2)	Fertilizer flow rate/(g/s)	Speed/ (m/s)	Weighting/ (g)	Mestimated/ (g)	PE/ (%)
SF	35	22.4	L	1446.3	1495.7	3.42
		28	M	1375.1	1437.4	4.53
		33.6	H	1437.5	1511.6	4.9
	55	35.2	L	2276.1	2216.7	2.61
		44	M	2256.8	2337.5	3.58
		52.8	H	2265.4	2364.7	4.38
DF	140	89.6	L	5508.5	5699.6	3.47
		112	M	5639.6	5895.4	4.54
		134.4	H	5693.3	5979.6	5.03
	200	128	L	8155.6	8456.5	3.69
		160	M	8185.3	8565.2	4.64
		192	H	8206.7	8620.4	5.04

Fig. 7.13 Centrifugal fertilizer spreader field discharge flow test: (a) Fertilizer tank. (b) Fertilizer discharging mechanism. (c) LED lights. (d) Terminal. (e) Jeston Nano. (f) Camera

vibrations accelerating fertilizer discharge [23]. Overall, the flow detection error remained within acceptable limits.

$$q = \frac{v \cdot B \cdot K}{20} \tag{7.9}$$

Where q is the fertilizer discharge flow rate per unit time of a single fertilizer discharge port, g/s; B is the fertilizer spreading width, m; K is the amount of fertilizer applied per unit area of the target plot, kg/hm^2; v is the real-time speed of the machine, m/s.

7.4 Conclusion

To address the challenge of accurately detecting the discharge flow rate of centrifugal variable-rate fertilizer spreaders, this study developed a centrifugal fertilizer discharge flow detection system based on an improved YOLOv5s-seg model. The system is portable and designed to be mounted on centrifugal variable-rate fertilizer spreaders [24]. Using a high-speed camera and a timed-interval frame method, it rapidly captures images of fertilizer particles and detects the fertilizer discharge flow rate in real time under different discharge port openings.

The improved YOLOv5s-seg model has a size of 6.66 MB, an FPS of 45 frames/s, a computational demand of 17.8 G, and an mAP of 95.9%. Compared with existing YOLOv5 series models, it effectively resolves issues such as missed detection of small particles, misdetection of overlapping particles, and inaccurate particle mask segmentation. It exhibits fast convergence speed, stable performance, and a lightweight architecture—making it suitable for deployment on mobile devices and capable of meeting real-time detection requirements. For granular fertilizers in non-overlapping, partially overlapping, and densely overlapping states, the total mass of particles counted per second by the detection system showed a strong linear relationship, indicating stable fertilizer discharge. At nine different discharge port openings (17, 18, 19, 20, 21, 22, 23, 24, and 25 mm), the coefficient of determination (R^2) between the fertilizer discharge flow rates measured by the imaging method and the weighing method reached 0.9984. For the high-flow dual-camera detection system, the correlation coefficient between the flow rates obtained by the two methods was 0.9997, demonstrating high detection accuracy. Field tests were conducted on the fertilizer spreader's discharge flow rate under different target application rates; the results showed that the maximum error of the fertilizer flow detection system was less than 4.9% (SF mode) and 5.04% (DF mode), respectively. However, the system has certain limitations: during high-speed field operation, shaking of the spreader may cause unstable fertilizer discharge; additionally, highly overlapping rounded particle contours can lead to misdetection of individual particles. Both factors may contribute to detection errors.

Overall, this study confirms that the proposed centrifugal variable-rate fertilizer discharge detection system can accurately measure the fertilizer discharge flow rate. Future work will focus on integrating real-time fertilizer flow rate data with the application control mechanism. This integration will enable precise, online adjustment of the discharge flow rate based on the spreader's operating speed and the fertilizer requirements of target plots, ultimately forming a closed-loop control system for fertilizer discharge flow.

References

1. Bagheri A, Tarighi J, Emami N et al (2024) Extension experts' intentions to use precision agricultural technologies, a test with the technology acceptance model [J]. Acta Technologica Agriculturae 27:84–91
2. Bovo E, Sorgato M, Lucchetta G (2023) An image-based approach for the mass flow measurement of plastic granules as an alternative solution to loss-in-weight feeding systems [J]. Powder Technol 430:119044
3. Cointault F, Sarrazin P, Paindavoine M (2003) Measurement of the motion of fertilizer particles leaving a centrifugal spreader using a fast imaging system [J]. Precision Agric 4:279–295
4. Feng X, Zhao K, Zhang J et al (2024) Carbonate reservoir fracture-cavity system identification based on the improved YOLOv5s deep learning algorithm [J]. Energy Sci Eng 12:2643–2660
5. Fu L, Yang Z, Wu F et al (2022) YOLO-Banana: a lightweight neural network for rapid detection of banana bunches and stalks in the natural environment [J]. Agronomy 12:391
6. Ghodki BM, Patel M, Namdeo R et al (2019) Calibration of discrete element model parameters: soybeans[J]. Comput Meth Biomech Biomed Eng: Part B 6:3–10
7. Huang J, Yang L, Zhang J et al (2021) Granular fertilizer mass flow measurement and vehicle experiments based on microwave Doppler method [J]. Appl Eng Agric 37:135–147
8. Jafari M, Hemmat A, Sadeghi M (2010) Development and performance assessment of a DC electric variable-rate controller for use on grain drills[J]. Comput Elect Agric 73:56–65
9. Janda A, Harich R, Zuriguel I et al (2009) Flow-rate fluctuations in the outpouring of grains from a two-dimensional silo[J]. Phys Rev E 79:031302
10. Janda A, Zuriguel I, Maza D (2012) Erratum: Flow rate of particles through apertures obtained from self-similar density and velocity profiles [Phys. Rev. Lett. 108, 248001 (2012)] [J]. Phys Rev Lett 109:189901
11. Jiang M, Liu C, Du X et al (2022) Flow detection mechanism of granular fertilizer based on light blocking principle [J]. Trans Chin Soc Agric Mach 53:91–99
12. Kollu PK, Bangare ML, Hari Prasad PV et al (2023) Internet of things driven multilinear regression technique for fertilizer recommendation for precision agriculture [J]. SN Appl Sci 5:264
13. Lei X, Wu W, Deng X et al (2023) Determination of material and interaction properties of granular fertilizer particles using DEM simulation and bench testing [J]. Agriculture 13:1704
14. Liang C, Yan Z, Ren M et al (2023) Improved YOLOv5 infrared tank target detection method under ground background [J]. Sci Rep 13:6269
15. Liu H, Wang L, Shi Y et al (2023a) A deep learning-based method for detecting granular fertilizer deposition distribution patterns in centrifugal variable-rate spreader fertilization [J]. Comput Elect Agric 212:108107
16. Liu Q, Gong X, Li J et al (2023b) A multitask model for realtime fish detection and segmentation based on YOLOv5[J]. PeerJ Comput Sci 9:e1262
17. Malta A, Mendes M, Farinha T (2021) Augmented reality maintenance assistant using YOLOv5 [J]. Appli Sci 11:4758
18. Pellicer LFAO, Ferreira TM, Costa AHR (2023) Data augmentation techniques in natural language processing [J]. Appl Soft Comput 132:109803
19. Qian L, Zheng Y, Cao J et al (2024) Lightweight ship target detection algorithm based on improved YOLOv5s [J]. J Real-Time Image Process 21:3
20. Reyes JF, Esquivel W, Cifuentes D et al (2015) Field testing of an automatic control system for variable rate fertilizer application [J]. Comput Elect Agric 113:260–265
21. Back S, Yu S, Kim Y et al (2014) An image-based application rate measurement system for a granular fertilizer applicator [J]. Trans ASABE:679–687
22. Singh P, Garg C, Namdeo A (2020) Applying machine learning techniques to extract dosages of fertilizers for precision agriculture [J]. IOP Conf Ser: Earth Environ Sci 614:012136
23. Vanbergeijk J, Goense D, Vanwilligenburg L et al (2001) PA—Precision Agriculture dynamic weighing for accurate fertilizer application and monitoring [J]. J Agric Eng Res 80:25–35

24. Wang B, Wang Y, Wang H et al (2022) Research on accurate perception and control system of fertilization amount for corn fertilization planter [J]. Front Plant Sci:13
25. Wang G, Li Q, Wang N et al (2023) SAFPN: a full semantic feature pyramid network for object detection [J]. Pattern Anal Appl 26:1729–1739
26. Wang S, Jiang H, Li Z et al (2024) PHSI-RTDETR: a lightweight infrared small target detection algorithm based on UAV aerial photography [J]. Drones 8:240
27. Wei L, Jin J, Deng K et al (2024) Insulator defect detection in transmission line based on an improved lightweight YOLOv5s algorithm [J]. Electr Pow Syst Res 233:110464
28. Yao S, Guan R, Huang X et al (2024) Radar-Camera fusion for object detection and semantic segmentation in autonomous driving: a comprehensive review [J]. IEEE Trans Intell Vehicle 9:2094–2128
29. Zhang J, Li X, Li J et al (2023) Rethinking mobile block for efficient attention-based models [C]. In: 2023 IEEE/CVF International Conference on Computer Vision (ICCV). IEEE, pp 1389–1400
30. Zhou X, Jiang L, Hu C et al (2022) YOLO-SASE: an improved YOLO algorithm for the small targets detection in complex backgrounds [J]. Sensors 22:4600

Chapter 8
Field Evaluation and Efficiency Analysis of Centrifugal Variable-Rate Fertilization in Precision Agriculture

Yinyan Shi, Xiaochan Wang, Z. Zhao, Xuekai Huang, S. Hu, and Man Chen

Abstract This study aims to investigate the economic benefits of centrifugal VRF technology in precision agriculture. To this end, a specialized VRF fertilization device for rice and wheat was developed and field trials were conducted in the rice-wheat rotation zone of Dongtai City, Jiangsu Province, China. The Nanjing 9108 rice variety was selected as the research subject, with three methods—manual broadcast fertilization, conventional empirical balanced fertilization, and real-time VRF fertilization—compared. Data were continuously collected for 3 years to conduct systematic analysis. From the perspectives of total fertilizer application, crop growth, and yield, real-time VRF technology reduced the annual average fertilizer application to 47.64 kg/ha, nearly 15.88% less than traditional empirical fertilization methods. With a slight reduction in nitrogen application, VRF technology increased rice yield to 8996.70 kg/ha, a 5.93% improvement over traditional cultivation methods. This figure effectively enhanced nutrient utilization and significantly improved crop spatial distribution. Considering economic benefits, it saved 95.28 yuan/ha in chemical fertilizer costs and increased net rice production revenue

Y. Shi (✉) · X. Wang
College of Engineering, Nanjing Agricultural University, Nanjing, China

Jiangsu Province Engineering Lab for Modern Facility Agriculture Technology & Equipment, Nanjing, China
e-mail: shiyinyan@njau.edu.cn; wangxiaochan@njau.edu.cn

Z. Zhao
College of Information and Electrical Engineering, China Agricultural University, Beijing, China
e-mail: zhaozhangcau@cau.edu.cn

X. Huang · S. Hu
College of Engineering, Nanjing Agricultural University, Nanjing, China
e-mail: 2022212011@stu.njau.edu.cn; 2023112024@stu.njau.edu.cn

M. Chen
Nanjing Research Institute for Agricultural Mechanization, Ministry of Agriculture and Rural Affairs, Nanjing, China
e-mail: chenman@caas.cn

Y. Shi et al. (eds.), *Technological Progress on Variable Rate Fertilization*, Smart Agriculture 17, https://doi.org/10.1007/978-981-95-4330-4_8

to 1007.62 yuan/ha. To comprehensively evaluate the performance of the fertilization device, field trial research plans were developed based on ISO5690 and ASAE341.2 standards. When nitrogen fertilizer application rates were set at 225 kg/ha, 300 kg/ha, and 375 kg/ha respectively, effective laying widths were measured as 21.05 m, 22.58 m, and 23.67 m. The coefficients of variation (CV) within particle size ranges were 11.51%, 9.25%, and 11.28% respectively. The average deviation of the overlapping layering experiment was 4.54%, with a coefficient of variation (CV) of 11.94%, fully meeting current technical standards for lateral distribution. Experimental results demonstrate that this centrifugal variable fertilization device significantly reduces chemical fertilizer usage, enhances crop yields, and improves spatial distribution consistency in farmland. The overall system meets operational requirements, providing theoretical foundations and practical references for promoting precision agriculture equipment.

Keywords Precision agriculture · Variable-rate fertilization · Centrifugal spreader · Field performance · Economic analysis

8.1 Introduction

Rice is one of the most important food crops in the world and has always played an important role in the domestic and foreign food supply system. Jiangsu is located in the center of the southern rice-growing area and generally adopts the rice-wheat rotation system. Under such circumstances, it is urgent to systematically improve the potential of grain yield increase and comprehensively improve the production capacity of food crops [1–3].

Fertilizers are indispensable for agricultural production, contributing significantly to boosting rice yields and safeguarding national food security [4]. However, the blind pursuit of high yields while disregarding fundamental fertilization principles leads to reduced nutrient absorption efficiency and low fertilizer utilization rates. This not only causes resource waste and diminished economic returns but also disrupts the balance of agricultural ecosystems, exerts adverse ecological impacts, and ultimately impedes the sustainable development of grain production [5, 6].

Scientific fertilization strategies and regionalized nutrient management are critical agronomic practices for ensuring high rice yields. These measures provide key technical support for yield improvement and lay the foundation for achieving grain production targets [7, 8]. Mechanized variable-rate fertilization (VRF) technology, a core component of precision agriculture, aligns with modern green farming concepts. By enabling precise nutrient application and balanced allocation, this technology substantially enhances resource utilization efficiency and markedly improves agricultural productivity.

As a cornerstone of precision agriculture, VRF technology leverages modern information systems to precisely match crop nutrient demands, thereby boosting agricultural productivity. Globally, it plays a vital role in mitigating ecological degradation caused by excessive fertilization, conforming to the strategic shift of modern agriculture toward green and sustainable development [10]. The "3S" technology—integrating Geographic Information Systems (GIS), Global Positioning Systems (GPS), remote sensing, and smart sensors—facilitates real-time monitoring of soil nutrient levels and crop growth dynamics. Through optimized targeted fertilization models and expert decision-support systems, intelligent agricultural equipment can now execute precise, scheduled fertilization operations across designated areas [11].

With the ongoing advancement of large-scale agricultural cultivation worldwide, centrifugal fertilizer spreaders have emerged as essential equipment in precision agriculture. Their compact design, moderate working width, high operational efficiency, minimal material loss, and cost-effectiveness have established them as core devices driving the industry's transition toward efficiency and intelligentization. These technological innovations have attracted considerable attention from agricultural research institutions, which now prioritize them as key research focuses [12, 13].

Regarding the structural design of centrifugal variable-rate fertilizer spreading equipment: Fulton et al. (Auburn University, USA) analyzed the physical properties of organic fertilizers and the fertilizer movement process via 3D model simulation. They designed an organic fertilizer variable-rate spreader that employs a hydraulic proportional valve to control the discharge opening, and compared the performance of the closed-loop control system (CLS) under different fertilizer application rate conditions. Tests revealed that the disc speed control accuracy ranged from 1–6 rpm, and the consistency error coefficients of variation (CVs) for fertilizer application were between 23 and 27.6% [14]. Cheol-woo et al. developed a centrifugal variable-rate fertilizer spreader with a uniform spreading mode by upgrading the controller and electric actuator to regulate the direction and quantity of fertilizer discharge. Test results demonstrated that uniform fertilizer application rates during variable-rate operation could be ensured by traveling adjacent passes in the same direction. Yang et al. designed a handheld centrifugal variable-rate fertilizer spreader for gardens, addressing the narrow-row spacing requirements of tea plantation operations. Granular fertilizer is discharged through a specially designed hoof-like outlet to control flow rate, and a DC motor drives the spreading disc to evenly disperse the fertilizer. Based on the current practical conditions of mechanized rice fertilization, Chen et al. developed a self-propelled variable-rate fertilizer spreader suitable for upland gaps in paddy fields. The system integrates sensor speed measurement signals and the corresponding fertilizer spreading width to generate control signal outputs, regulating the speeds of a stepper motor and hydraulic motor to adjust the fertilizer opening and disc speed, thereby enabling continuous variable-rate fertilization. Test results showed the spreader had a fertilizer application rate deviation of 5%, a fertilizer discharge stability coefficient of 6.2%, and a working efficiency of 6 hm^2/h—20%–50% higher than that of other spreaders [15–17].

In terms of performance optimization technology for centrifugal variable-rate fertilizer spreading equipment: Liedekerke, Coetzee et al. calculated the motion trajectory and dynamic model of granular fertilizer on the disc and in flight for a centrifugal variable-rate spreader, and established a discrete element method (DEM) simulation model for the disc spreader. They analyzed the influence of fertilizer particle physical parameters (particle size, friction coefficient, restitution coefficient) on fertilizer distribution patterns, and verified the simulation model's accuracy and feasibility through experiments. Kweon et al. (University of Illinois, USA) conducted a series of studies on the fertilizer outlet gate of a self-designed single-disc variable-rate spreader, theoretically analyzing the feed gate adaptive algorithm. By setting different area, shape, and position parameters for the drop gate, they performed simulated fertilization tests and confirmed that this algorithm could adapt to specific application rate requirements and achieve high-quality fertilizer distribution. Lv et al. designed a cone-disc fertilizer spreading device; based on an analysis of the fertilizer particle motion model, they optimized the device structure via bench tests to meet potato fertilization needs. To improve the spreading accuracy and fertilizer distribution uniformity of a self-developed centrifugal variable-rate spreader, Shi et al. conducted DEM simulation tests of spreading performance using common compound granular fertilizers as test materials, investigating the relationship between particle distribution variation rules and spreader operating parameters. Field validation tests showed that the mean CV of particle distribution within the spreader's effective width was 16.74%, complying with relevant standard operating requirements for centrifugal fertilizer spreaders [18, 19].

In recent years, with the large-scale development of modern agricultural cultivation and the vigorous implementation of China's national "double reduction" plan, fertilization efficiency and fertilizer utilization have garnered increasing attention. Small-width outer groove wheel-type variable fertilizer applicators have gradually been replaced by high-efficiency centrifugal variable-rate spreaders, making centrifugal variable spreading technology one of the development directions for advancing high-efficiency precision agriculture [20, 21]. Professor Ehlert's team developed a nitrogen VRF spreader for winter wheat based on a mechanical pendulum-type crop density detector; the on-board computer continuously adjusted the spreader's application rate according to real-time fixed-point crop density measurements. Multiple field experiments demonstrated that the spreader reduced fertilizer use by 10–12% while improving crop yield and quality [22]. Karl-Heinz et al. developed a fungicide variable sprayer based on cereal leaf area index (LAI); a portable spectral sensor (CROP-Meter) measured crop biomass density in real time, matched with a crop canopy LAI model, and combined with GPS positioning information to output fungicide dose control signals, driving a hydraulic valve to adjust fungicide flow for variable spraying [23, 24]. Professor Wang's research group designed a liquid fertilizer variable deep application machine based on prescription maps to improve crop fertilizer utilization. The lower computer receives photoelectric sensor signals and BeiDou satellite position data to control an electric pressure-regulating valve for target application rate adjustment [25, 26]. The machine's system variable fertilization error was <5.08%, and it increased yield per mu by approximately 13.2%. Zhang et al. developed a chain conveyor variable fertilizer spreader based on prescription map information using VRF technology; the maximum error between actual and preset application rates was 7.53%, the spreading width reached 30 m, and the CV of fertilization within the 16 m effective width was 14.77%—four times higher than that of fixed-width variable-rate fertilization, meeting practical production requirements. Based on current conditions of mechanized rice fertilization, Chen et al. designed a self-propelled variable-rate spreader suitable for high ground gaps in paddy fields, enabling continuous variable spreading. It achieved a fertilizer application deviation of 5%, a fertilizer stability coefficient of 6.2%, a uniformity CV of 14%, a spreading width of 14 m, and a working efficiency of 6 hm^2/h—20%–50% higher than other spreaders. Shi et al. developed a sensor-based double-disc centrifugal spreader for rice surface VRF, which integrates expert fertilization strategies to respond to target application rates, adjusts real-time crop fertilization rates online, achieves relatively precise rice VRF, and improves fertilizer distribution uniformity and application rate accuracy [27, 28].

A review of existing research on VRF technology reveals a primary focus on the development of key technologies for VRF equipment and verification experiments based on prescription map information or real-time sensors [29]. However, in recent years, research on the impacts of VRF technology on crop growth and economic benefits—particularly evaluations of the operational effectiveness of independently developed VRF equipment—has been rarely reported [30]. Consequently, the balanced distribution of fertilizer nutrients and the efficiency of fertilizer resource utilization have become focal points for researchers [31]. There is an urgent need to explore a harmonious and stable sustainable development pathway that can enhance grain yield, ensure food production security, minimize environmental and resource pollution, and maximize reductions in economic losses from fertilizer waste [32].

Therefore, this study focused on a centrifugal variable-rate fertilizer spreader independently developed by our research group, which is based on real-time crop growth information. Using compound granular fertilizers commonly applied in the rice-wheat rotation areas of Jiangsu as test materials, we conducted performance tests of the spreader under multi-trajectory overlapping fertilization operations. The objectives were to explore the formation mechanism and distribution laws of fertilizer particles for the multi-trajectory variable spreader under different working parameters, operating conditions, and travel modes, and to evaluate the overall machine's working effectiveness in multi-trajectory overlapping fertilization. This research aims to improve the particle distribution uniformity and fertilization accuracy of dual-disc centrifugal variable-rate spreaders, mitigate the issues associated with centrifugal spreading of granular fertilizers, and better align with the expectations of farmers implementing precision agriculture VRF technology and government departments promoting intelligent management strategies—such as nutrient site-specific management (SSM) and global navigation satellite systems (GNSS).

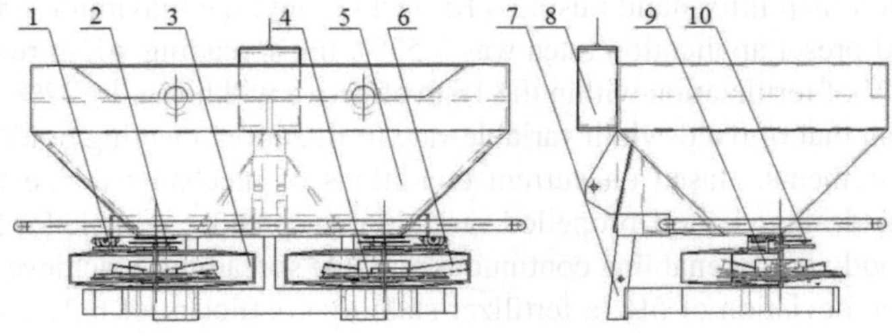

(a) Main view (b) Left view

1. Electric control regulating device 2. Fertilizer discharge port 3. Fertilizer stopper plate (cover) 4. Spreading disc 5. Fertilizer tank 6. Fertilizer paddle blade 7. Frame 8. Control box 9. Fertilizer equating hood 10. Manual regulating device

Fig. 8.1 Structural diagram of variable-rate fertilizer spreading device. (**a**) Main view. (**b**) Left view. 1 Electric control regulating device. 2 Fertilizer discharge port. 3 Fertilizer stopper plate (cover). 4 Spreading disc. 5 Fertilizer tank. 6 Fertilizer paddle blade. 7 Frame. 8 Control box. 9 Fertilizer equating hood. 10 Manual regulating device

8.2 Materials and Methods

8.2.1 Test Equipment

8.2.1.1 Variable-Rate Fertilization Equipment and Operating Principle

The fertilization and spreading test apparatus, independently developed by our research team (Fig. 8.1), utilizes a dual-disc centrifugal adjustable fertilizer distribution system. Its design philosophy focuses on real-time collection and analysis of crop growth status (spectral characteristics) data. The complete system comprises a frame structure, fertilizer storage tank, discharge mechanism, adjustment device, spreading disc, flow guide plates, transmission system, and control system.

The tractor's three-point suspension system is connected to a fertilizer application unit consisting of two independent tanks. Each tank is equipped with a filtration unit and a mixing mechanism, while a dual-channel fertilizer discharge device at the bottom enables simultaneous multi-point distribution. The discharge rate is precisely regulated by a stepper motor that adjusts the opening of the discharge port. Positioned directly beneath the outlet, the fertilizer spreader generates centrifugal force during high-speed operation to disperse nutrients evenly within the constant-speed hood. Driven by the tractor's power system, a pair of high-speed rotating spreaders propel the fertilizer into the target area, achieving uniform coverage of single-row fields with widths exceeding 24 m.

The fertilizer application system comprises a fixed base plate and two adjustable side plates. The movable plates are designed with a fan-shaped opening, which collaborates with the fixed plate to control fertilizer discharge and realize dynamic adjustment of the application rate. Detailed structural components are depicted in Fig. 8.2.

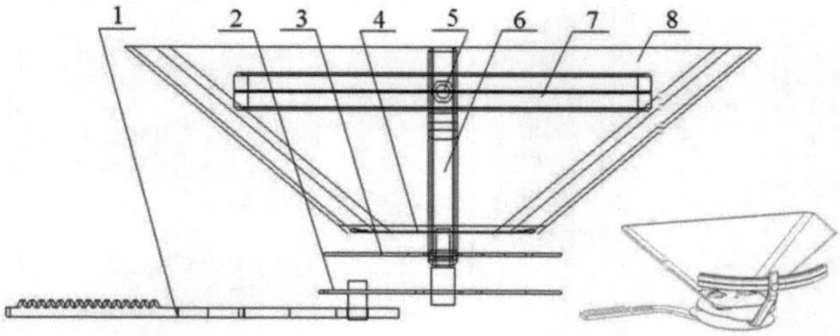

Fig. 8.2 Structural diagram of fertilizer distributing device. 1 Automatic control lever. 2 Dynamic fertilizer disc 1. 3 Dynamic fertilizer disc 2. 4 Fixed fertilizer disc. 5 Setting screw. 6 Manual adjusting handle. 7 Scale line. 8 Fertilizer tank

In Fig. 8.2, Dynamic Fertilizer Discharge Disc 1 functions as the automatic control adjustment disc, while Dynamic Fertilizer Discharge Disc 2 serves as the manual control adjustment disc. Under automatic control mode, the feeding port of Dynamic Fertilizer Disc 2 is adjusted to its maximum opening, and the fertilizer dosage is regulated by controlling the relative position of the discharge port between Dynamic Fertilizer Disc 1 and the fixed fertilizer disc. The variable-rate fertilizer spreader adopts two independent left and right fertilizer discharge channels operating simultaneously. Neglecting the influence of external factors such as the operating environment, the model for the fertilizer discharge flow rate per unit time at the feeding port is as follows:

$$q = \frac{v \cdot B \cdot Q \times 10^3}{2 \times 10^4} = \frac{v \cdot B \cdot Q}{20} \tag{8.1}$$

where, q is the fertilizer discharge flow rate per unit time of a single discharge port, g/s; B is the width of fertilizer spreading, m; Q is the target fertilizer amount per unit area, kg/hm2; v is the travel speed of the whole machine, m/s.

8.2.1.2 Structure and Working Principle of the Real-Time Variable-Rate Fertilizer Spreader

The real-time variable-rate fertilization field test platform developed in this study takes the centrifugal variable-rate fertilization device—independently developed by our research team—as its core component. The system achieves precise fertilization by leveraging rice growth data. The overall structural design can be referenced from the literature [23]. Its main components include the tractor's rear three-point suspension system and a pair of counter-rotating fertilizing discs driven by a power output shaft (PTO), with a single operation covering an effective width exceeding 24 m. The working principle is illustrated in Fig. 8.3.

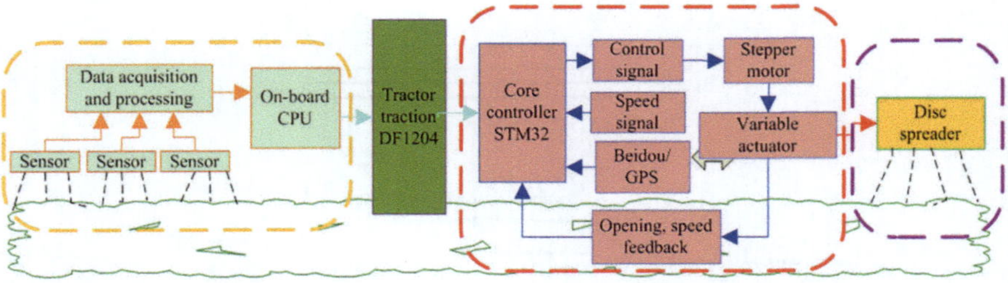

Fig. 8.3 Schematic diagram of variable rate fertilization

During field fertilization operations, spectral detection equipment collects normalized differential vegetation index (NDVI) data from rice canopies. This data is transmitted to the on-board control terminal via a wireless communication module, requiring real-time processing and analysis. To address this demand, our research team has developed an intelligent variable fertilization decision-making system, which is designed to accurately calculate target application rates. Built on a vehicle-mounted computer platform, the system integrates real-time feedback parameters including vehicle speed, disc rotation frequency, and fertilizer box opening degree. Guided by expert system recommendations, the core controller continuously adjusts the crop fertilization rate, realizing precise variable control in rice fertilization.

8.2.2 *Test Conditions*

The test material was compound fertilizer (N-P_2O_5-K_2O, 30-10-6)—a type commonly used in the rice-wheat rotation areas of Jiangsu Province in the middle and lower reaches of the Yangtze River. The fertilizer particles were simplified and equated to spherical particles, and their physical property parameters were determined via random sampling. The particles had an average diameter of 4.02 mm, a sphericity of >0.94, an average moisture content of 16.42%, a density of 1.38×10^3 kg·m^{-3}, a hardness of 44.73 N·mm^2, a kinetic friction coefficient of 0.26, a static friction coefficient of 0.34, a Poisson's ratio of 0.24, and a shear modulus of 10.76 MPa.

The experiment was conducted in May 2022 at the test site of Jiangsu Jinmansui Agricultural Development Co., Ltd. (Dongtai, Jiangsu, China) under sunny weather conditions. The environmental parameters were as follows: temperature 15–25 °C, wind speed <2.0 m/s (complying with the permissible test wind speed specified in ASAE standards), relatively flat terrain (slope < 2%), test plot area 500 m^2, and air relative humidity 62%.

8.2.3 *Test Methods*

8.2.3.1 Fertilizer Spreading Test Methods and Field Implementation

To ensure standardized operations and data accuracy in the multi-pass overlapping application test of the centrifugal variable-rate fertilizer spreader, the experiment was performed in accordance with the test standards specified in GB/T 5262-2008 *General Rules for Determining Test Conditions of Agricultural Machinery*, ISO 5690, and ASAE S314.2. The performance test of the spreader was carried out following the schematic diagram shown in Fig. 8.4.

Plastic film was laid in the test area (45 m × 20 m) to reduce fertilizer waste, and an appropriate amount of soft cotton fabric was attached to the inner wall of each collection pan to minimize test errors caused by fertilizer bouncing. A total of 253 collection pans (size: 50.8 × 40.6 × 10.2 cm) were arranged in a 23 × 11 (column × row) matrix, with a column spacing of 2 m and a row spacing of 1 m. During the fertilization test, the corresponding operating parameters were adjusted. The traction tractor (Dongfeng 1204) traveled through the central path at a speed of 1.2 m/s for a single-pass fertilization operation; subsequent adjacent passes were spaced at different distances (corresponding to the effective operating width), and multi-pass overlapping fertilization was conducted using two operating modes: reciprocating travel (reverse direction) and rotary travel (same direction). After the fertilization operation was completed, the mass of fertilizer particles in each collection pan was weighed (and stored in labeled bags), which was used to characterize the transverse distribution of fertilization by the variable-rate fertilizer applicator. For each test, the fertilizer box was filled to 50% of its total capacity, and the machine was calibrated to the corresponding working state. The variable-rate fertilizer spreader was started in the stabilization zone before entering the measurement zone (the area with collection pans), and a stopping zone was set to ensure the test accuracy of the measurement zone. The coefficient of variation (CV) of fertilizer particle distribution within the effective fertilization width was calculated separately to evaluate the fertilization performance. The travel trajectory and collection pan array are shown in

Fig. 8.4 Performance test of spreading fertilizer

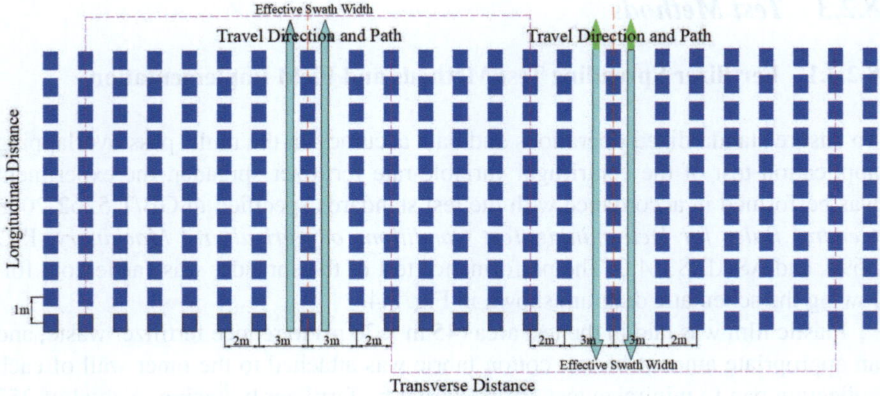

Fig. 8.5 Schematic diagram of collection pan matrix and fertilizer spreader track

Fig. 8.5, and each group of experiments was repeated three times. The formula for calculating the CV of fertilizer particle distribution is as follows:

$$CV = \frac{SD}{\overline{X}} \times 100\%$$

$$SD = \left\{ \sum_{i=1}^{n} \left[\left(X_i - \overline{X} \right)^2 \right] / (n-1) \right\}^{\frac{1}{2}} \tag{8.2}$$

$$\overline{X} = \frac{1}{n} \sum_{i=1}^{n} X_i \qquad (i = 1,2,3,\ldots,n)$$

Where, *CV* is the coefficient of variation of particle distribution; *SD* is the standard deviation of fertilizer particle mass; $\overline{X}$ is the mean value of fertilizer particle mass; X_i is the mass of fertilizer particles in the *i*th collection pan, g; *n* is the number of collection pans.

8.2.3.2 Performance Testing Protocol and Experimental Field Design

According to the current cultivation plan of Jinmansui Rice Cooperative, each experimental plot covers an area of 1.333 hm^2 (256 m in length and 50 m in width). This study selected nine plots as research samples, categorized as follows: the manual fertilization group (plots FA-01, FA-02, FA-03), the traditional mechanical fertilization group (plots FB-01, FB-02, FB-03), and the real-time variable-frequency mechanical fertilization group (plots FV-01, FV-02, FV-03). All plots received standardized organic fertilizer application before basal fertilization. Rice seedling cultivation typically commenced between May 10th and 15th each year, with transplanting operations conducted from June 22nd to 27th. Subsequent fertilization

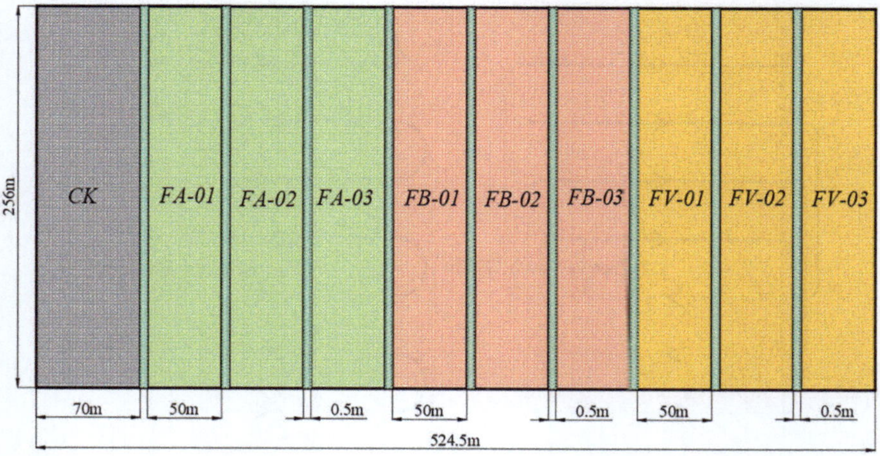

Fig. 8.6 Test field plots of fertilizer spreading

Artificial fertilizer spreading Traditional empirical spreading Real-time variable-rate spreading

Fig. 8.7 Three kinds of fertilizer spreading test

was implemented during the tillering stage (June 22nd–27th) and the flowering protection period (August 7th–12th). Rice harvesting was uniformly completed by combine harvesters from late October to early November.

The fertilization process is illustrated in Fig. 8.6. During the experimental phase, researchers established a control plot with two treatment groups: Group 1 applied standard fertilizer rates (300 kg/hm^2), while Group 2 utilized a custom centrifugal fertilization system. The system employs fixed flow rate control, maintaining an average operational speed of 1.2 m/s. Ground-mounted Hall sensors monitored the system's performance, whereas the self-developed variable-speed centrifugal fertilizer dynamically adjusted its discharge rate to ensure uniform distribution across the fertilized area. All equipment maintained a constant operating speed of 1.2 m/s throughout the process, with real-time monitoring through Hall sensors.

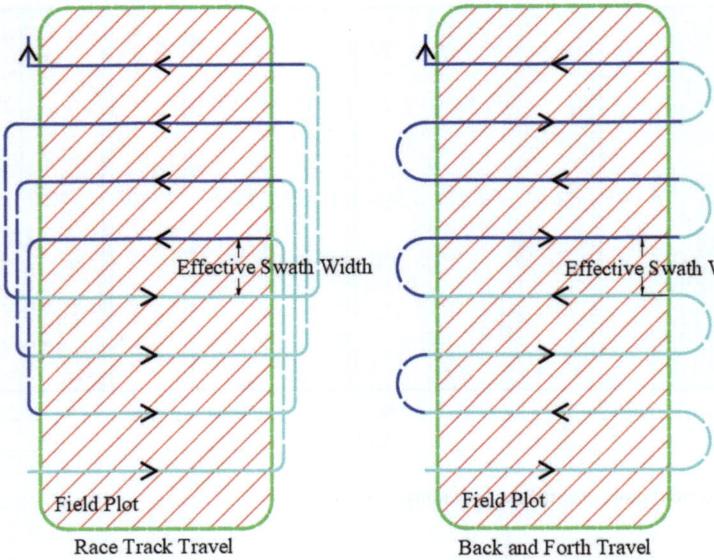

Fig. 8.8 Operating mode of fertilizer spreader travel

8.2.4 *Test Setup*

In practical large-scale field fertilization operations, fertilizer spreaders commonly adopt multi-pass adjacent travel to achieve overlapping fertilizer distribution. In accordance with the ASAE S314.2 standard, two travel path modes are specified for fertilizer spreaders: one with adjacent paths traveling in the same direction, and the other with adjacent paths traveling in opposite directions (as illustrated in Fig. 8.8). Co-directional travel, also termed race-track pattern fertilization or turn-back fertilization, refers to successive adjacent paths operating in the identical direction. Reverse travel, alternatively called the reciprocal operation mode, denotes consecutive adjacent paths operating in opposing traveling directions.

The track spacing between adjacent passes is determined by the effective width of single-pass fertilization, and two methods exist for defining this effective width. The first method is based on the particle deposition distribution pattern of single-pass fertilization (shown in Fig. 8.9, pink line). It identifies either the two positions on the left and right sides of the centerline (the fertilizer spreader's travel path) where the fertilizer amount equals half of the target amount at the centerline, or the interval between two points exhibiting the highest uniformity of fertilizer distribution. The second method relies on the coefficient of variation (CV) of particle distribution after overlapping fertilization across multiple adjacent passes (shown in Fig. 8.9, with blue, pink, and orange lines). Specifically, the effective spreading width is defined as the maximum fertilization width corresponding to the deposition pattern when the CV reaches its minimum value or falls within the allowable range

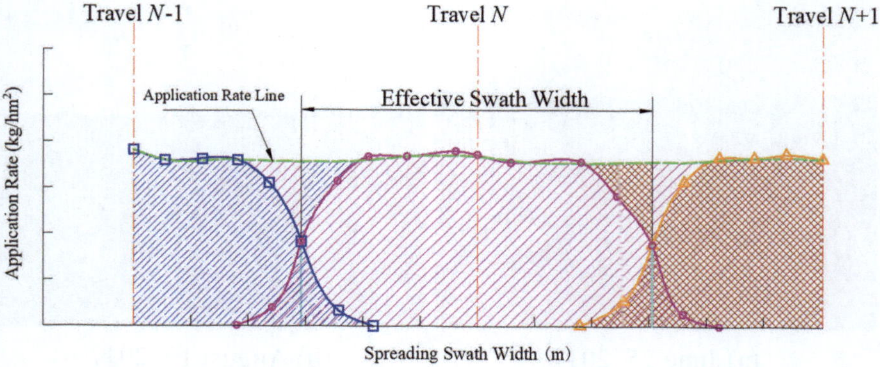

Fig. 8.9 Schematic diagram of the effective swath width and the particle distribution for multi-passes overlap

(typically CV < 15%). Beyond this effective width, the CV of fertilizer particle distribution increases as the spreading width expands.

8.2.4.1 Fertilizer Application Rate Measurement

In the real-time variable-rate spreading test area, the initial fertilizer amount (M_0) in the hopper was recorded before each single-pass operation. After completion, the remaining fertilizer amount (M_1) was measured, yielding the actual spread amount for that pass as $M_a = M_0 - M_1$. Concurrently, during each pass, a spectral sensor collected normalized differential vegetation index (NDVI) data from the rice canopy, which was stored in the on-board computer. The mean NDVI value (N_m) was calculated for subsequent analysis, and the corresponding target application rate (M_t) was determined using the expert decision system's fertilization model. The control precision of the variable-rate spreader was evaluated by comparing M_a with M_t. Additionally, the actual application rate of the variable-rate system was benchmarked against traditional empirical fertilization methods. The variable-rate test included three round trips, generating six sets of single-pass data.

8.2.4.2 Population Growth Determination

As shown in Fig. 8.5, hand-held spectral data collection was performed on the three groups of test areas, to which the same amounts of organic base fertilizer were applied. The NDVIs of the rice crop canopies before and after fertilization of the nine operation plots were collected in June and August of each year. Three sets of spectral data were selected for each control area. Based on a comparison of the NDVI values and variability of spectral data of the same small area (200 m^2)

(a) June 25, 2017 (b) August 10, 2017

Fig. 8.10 Test of spectral data acquisition. (**a**) June 25, 2017 (**b**) August 10, 2017

randomly selected from each plot, the population structure of rice crop growth was analyzed using average values. The coefficient of variation is given as follows:

$$\begin{cases} C_V = \dfrac{SD}{\overline{X}} \times 100\% \\ SD = \left\{ \sum_{i=1}^{n} \left[\left(X_i - \overline{X} \right)^2 \right] / (n-1) \right\}^{\frac{1}{2}} \\ \overline{X} = \dfrac{1}{n} \sum_{i=1}^{n} X_i \qquad (i = 1,2,3,\ldots,n) \end{cases} \tag{8.3}$$

where *Cv* is the coefficient of variation of the NDVI values, *SD* is the standard deviation reflecting the NDVI values, $\overline{X}$ is the average of the NDVI values, and X_i is the NDVI value in the *i*th test point, and *n* is the total number of test points.

8.2.4.3 Yield Measurement

Each November, rice harvests were completed using combine harvesters. Three representative plots were selected for yield assessment. Each plot was divided into six standardized 100 m^2 sections (10 m × 10 m) using diagonal sampling. Representative sample points were randomly chosen, and average weights per sampling unit were measured to calculate actual yield per hectare.

8.3 Results and Analysis

8.3.1 Application Rate Analysis of Variable-Rate Fertilization

In the real-time variable-rate fertilization zone, the newly-developed centrifugal variable-rate-rate fertilizer spreader was utilized to implement two core concepts: within-plot uniform fertilization tailored to actual demands and between-plot variable-rate fertilization based on real-time rice growth data. The fertilizer application rate results are presented in Table 8.1. Analysis of the differences between the target demand fertilizer application rate (Mt) and the actual spreading amount (Ma) across plots with varying rice growth conditions revealed minimal control errors in the variable-rate spreader system. Moreover, the control precision for each plot remained above 80%, with an average of 88.06%, a minimum of 81.68%, and a maximum of 96.09%. These findings indicate that the variable-rate system exhibits excellent control performance, ensuring reliable system control precision.

According to the fertilizer application rate test results presented in Table 8.1, the total fertilizer applied in the 6 groups of variable-rate fertilization trials in 2017 was 1018.37 kg, covering a total area of 4 hm^2 across 3 plots. In comparison, the local traditional empirical balanced fertilization rate was 300 kg/hm^2 (totaling 1200 kg). Consequently, this variable-rate spreading experiment achieved a reduction of approximately 45 kg/hm^2 in application rate, corresponding to a per-unit-area fertilizer saving of about 15.14%. For the same variable-rate fertilization test area, the fertilizer usage in 2018 was 1006.52 kg, representing a reduction of 48.37 kg/hm^2 compared to traditional empirical balanced fertilization. This translates to a per-unit-area fertilizer saving of approximately 16.12%. In 2019, the fertilizer used amounted to 1003.49 kg, a reduction of 49.13 kg/hm^2 relative to traditional empirical balanced fertilization, equating to a per-unit-area saving of 16.37%. From 2017 to 2019, the variable fertilizer application rate in each test area showed a year-on-year decrease, with an average application amount of 1009.46 kg. Compared to traditional empirical balanced fertilization, this resulted in an approximate

Table 8.1 Test results of application rate of variable-rate fertilizer (VRF)

Test		N_m	M_t/kg	M_a/kg	Control precision /%	Error /%
VRF	1	0.485	174.30	182.13	95.70	/
	2	0.423	186.49	154.65	82.93	
	3	0.398	193.01	173.37	89.82	
	4	0.502	169.69	140.06	82.14	
	5	0.467	185.24	151.32	81.68	
	6	0.346	208.38	216.84	96.09	
Total		Year 2017		1018.37	/	15.14
		Year 2018		1006.52		16.12
		Year 2019		1003.49		16.37
Average		/		1009.46		15.88
Empirical		20 × 10 × 6		1200		

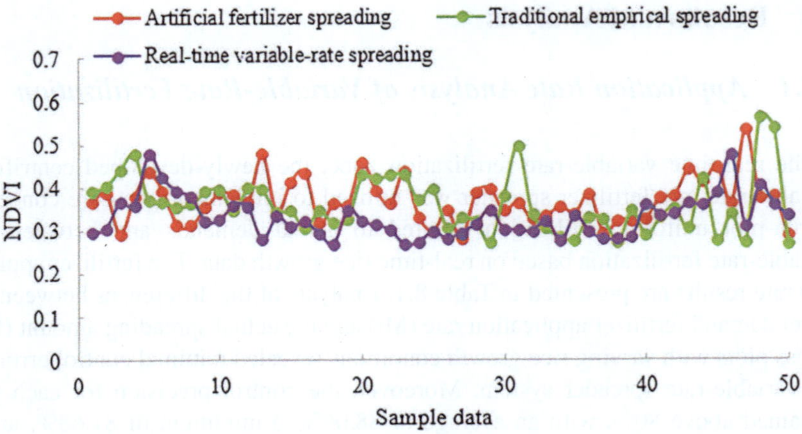

Fig. 8.11 NDVIs of crop spectral data before fertilizer

reduction of 47.64 kg/hm^2, corresponding to a per-unit-area fertilizer saving of about 15.88%. These results demonstrate that the newly developed variable-rate fertilizer spreader achieved significant real-time fertilizer savings and application rate reductions.

8.3.2 Analysis of Rice Population Growth Structure

Figure 8.11 displays the statistical distribution of normalized differential vegetation index (NDVI) values derived from rice spectral data across experimental plots prior to top-dressing. The results indicate no significant differences in growth patterns among the three experimental groups: the average canopy NDVI values were 0.383, 0.368, and 0.360, respectively; the maximum values reached 0.548, 0.531, and 0.562; and the minimum values were 0.258, 0.278, and 0.267. The coefficients of variation (CVs) for NDVI were 15.40%, 14.21%, and 14.57%, respectively. These findings confirm pronounced spatial heterogeneity in regional rice growth and uneven distribution of population structure, leading to distinct fertilizer requirements across the experimental plots. Thus, fertilization strategies should be adjusted in a timely manner based on crop growth dynamics during top-dressing operations.

Following fertilizer application on June 25, 2017, the research team used a green-light spectral sensor to acquire rice canopy NDVI data in the experimental fields on August 10 of the same year (see Fig. 8.11). As shown in Fig. 8.12, the fertilization treatments significantly improved rice growth status in the experimental areas, which was evidenced by the substantial increase in canopy NDVI values. The average NDVI values across different experimental areas were 0.575, 0.597, and 0.639; the maximum values reached 0.737, 0.714, and 0.706; and the minimum values were 0.314, 0.393, and 0.524. These data suggest that scientifically adjusted

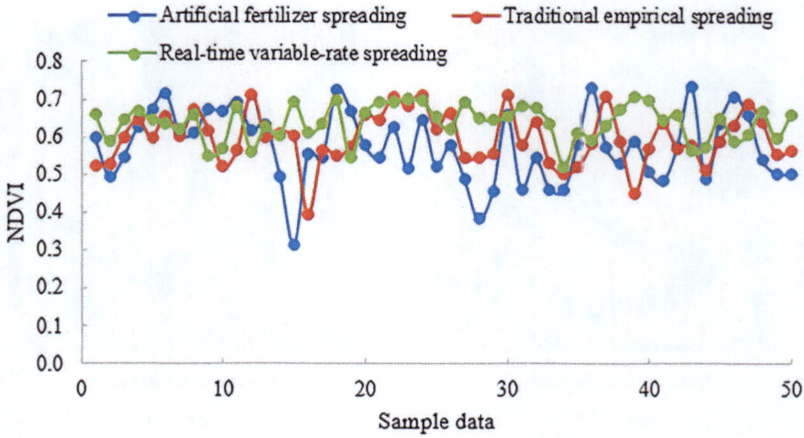

Fig. 8.12 NDVIs of crop spectral data after fertilizer

fertilization parameters are beneficial to natural rice growth. Statistical analysis revealed significant spatial heterogeneity in crop growth across the experimental areas, with NDVI CVs of 16.07%, 11.55%, and 7.22%, respectively. Compared with pre-fertilization conditions, the mechanical fertilization group exhibited more notable improvements in population structure, with its NDVI CV decreasing from 14.57% to 7.22%. In contrast, traditional mechanical fertilization methods failed to improve crop population structure. For the manual fertilization areas, the population structure tended to deteriorate before and after fertilization with the difference in spatial distribution increasing by 0.67 percentage points. The research indicated that implementing precision variable fertilization technology based on crop growth dynamics has high scientific validity and feasibility, whereas traditional uniform fertilization and manual application methods have obvious limitations. Their lack of precise guidance matching actual crop needs leads to population structure imbalance, resulting in resource waste and a significant reduction in fertilizer utilization efficiency.

From 2018 to 2019, the NDVIs of rice growth before the application of topdressing tillering fertilizer were obtained from 20 m × 20 m areas that were randomly selected in each plot. The data distribution of crop growth in the different operating areas is shown in Fig. 8.13.

In 2018, prior to fertilization, the average NDVI values of rice spectral data in the manual fertilization zone, traditional mechanical fertilization zone, and real-time variable fertilization zone were 0.363, 0.374, and 0.381, with CVs of 17.01%, 18.89%, and 12.78%, respectively. After fertilization, the NDVI values of these zones increased to 0.388, 0.382, and 0.387, with CVs of 19.36%, 19.82%, and 11.96%, respectively. Comparative analysis of NDVI data from different experimental plots between 2017 and 2019 showed consistent crop growth trends across all test areas. Analysis of NDVI CVs confirmed significant spatial heterogeneity in crop growth: the manual fertilization zones exhibited distinct plant distribution

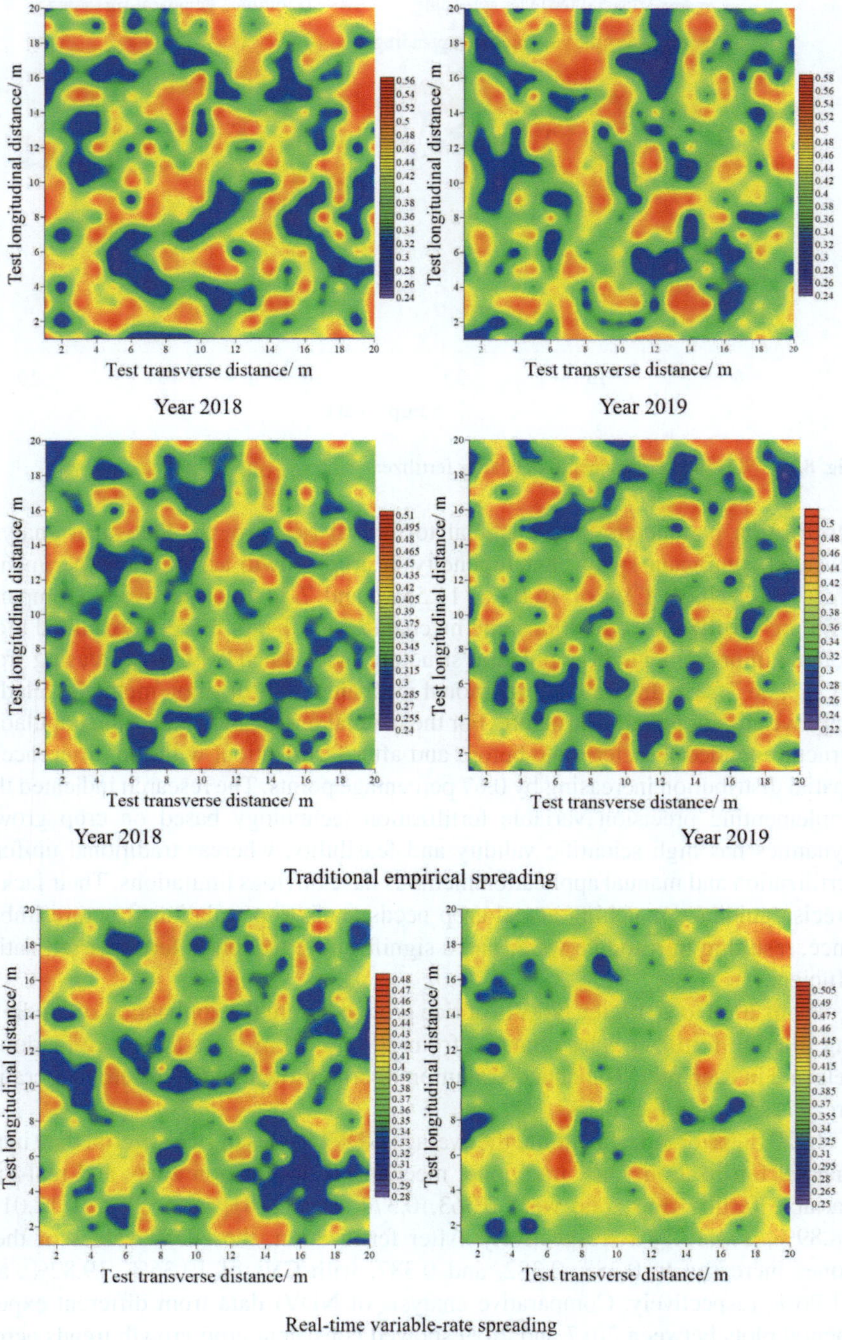

Fig. 8.13 NDVIs of crop spectral data in different plots before fertilizer during year 2018–2019

patterns compared to the traditional mechanical fertilization zones, with chaotic plant arrangements despite increased yields. In contrast, the real-time variable fertilization zones showed smaller spatial variations, with more rational and continuously optimized plant structures. These results demonstrate that the real-time variable fertilization method has clear advantages in enhancing the spatial uniformity of crop growth distribution.

Figure 8.14 presents the dynamic map of the net daily yield index after rice tillering fertilizer application, illustrating the distribution of crop growth data in each management area during 2018–2019.

In 2018, crop growth in the experimental zones showed an overall upward trend following rice sowing. The average normalized vegetation index (NDVI) was 0.561 in the manual fertilization area, 0.574 in the traditional mechanical broadcasting area, and 0.619 in the real-time variable fertilizer applicator area, with corresponding coefficients of variation (CVs) of 19.77%, 20.68%, and 8.89%. By 2019, after fertilization, the average NDVI values of these three areas had changed to 0.572, 0.582, and 0.621, respectively, and their CVs were further reduced to 20.28%, 21.54%, and 7.41%.

Based on the 2017–2019 fertilization treatment experimental data (field plot details provided in Table 8.2), this study revealed the spatial heterogeneity characteristics of crop growth. The NDVI values of rice in the experimental plots (real-time variable fertilization area) consistently exceeded those in the manual fertilization zone and traditional empirical fertilization area, indicating that precision variable fertilization combined with scientific soil nutrient management significantly promotes rice growth. Using NDVI CVs as an evaluation metric, both the manual fertilization zone and traditional empirical fertilization area exhibited chaotic crop community structures that deteriorated progressively over time. Although conventional mechanical broadcasting technology has clear advantages in terms of coverage area, it fails to achieve uniform spatial distribution and rational community structure in the field. In contrast, the application of intelligent fertilization equipment with real-time dynamic adjustment capabilities has created a more balanced growth environment and significantly improved the community structure in paddy fields. Research data shows that after precise regulation, the CV of rice spectral NDVI parameters remained stable within the range of 7.00%–9.00% for most of the time. This suggests that prolonged excessive or insufficient fertilization easily leads to the degradation of soil physicochemical properties, aggravates nutrient imbalance, and adversely affects crop growth and yield improvement. Therefore, establishing dynamic fertilization strategies tailored to actual crop needs and adjusting application rates in a timely manner are essential for effectively improving soil quality and achieving balanced nutrient management.

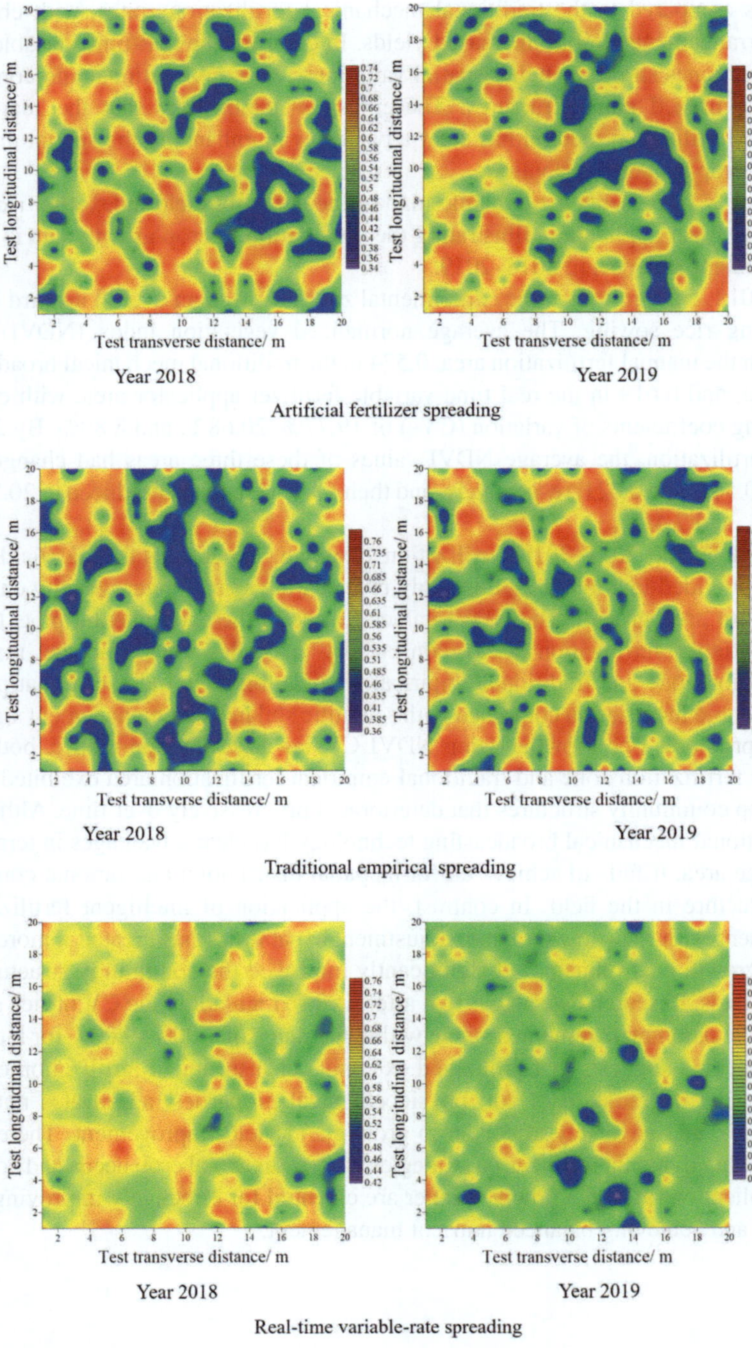

Fig. 8.14 NDVIs of crop spectral data in different plots after fertilizer during year 2018–2019

Table 8.2 Analysis of NDVI data in different plots before and after fertilizer during year 2018–2019

Item			*FA*	*FB*	*FV*
2017	Before fertilization	Mean NDVI	0.383	0.368	0.360
		Coefficient of variation	15.40%	14.21%	14.57%
	After fertilization,	Mean NDVI	0.575	0.597	0.639
		Coefficient of variation	16.07%	11.55%	7.22%
2018	Before fertilization	Mean NDVI	0.363	0.374	0.381
		Coefficient of variation	17.01%	18.89%	12.78%
	After fertilization,	Mean NDVI	0.561	0.574	0.619
		Coefficient of variation	19.77%	20.68%	5.59%
2019	Before fertilization	Mean NDVI	0.388	0.382	0.387
		Coefficient of variation	19.36%	19.82%	11.96%
	After fertilization,	Mean NDVI	0.572	0.582	0.621
		Coefficient of variation	20.28%	21.54%	7.41%

8.3.3 Analysis of Crop Yield Synergy

Table 8.3 shows the statistics of rice yield in each experimental area (artificial fertilizer spreading area, traditional empirical spreading area, and real-time variable-rate spreading area) from 2017 to 2019.

An analysis of the rice yield data for different operation plots (outlined in Table 8.3) showed that in 2017, the CVs of crop yields for the six test groups in the three experimental areas were 12.25%, 10.29%, and 6.13%, respectively. These results indicate that the spatial variation in rice yield was the largest in the manual fertilization area, followed by the traditional mechanical broadcasting area, while the real-time variable mechanical broadcasting area had relatively balanced spatial yield variation. This demonstrates that variable-rate fertilization improves the population structure of crop growth, provides a good foundation for crop photosynthesis, and establishes a sound framework for the final high yield of rice. This is consistent with the aforementioned analysis of NDVI data, which reflects the crop growth population structures in the experimental areas. Thus, the analysis of spatial differences in rice yield among the test plots for 2018 and 2019 will not be repeated herein.

In 2017, three rice cultivation models were tested in the experimental field: manual fertilization, traditional empirical mechanical sowing, and variable-controlled mechanical sowing, with average yields of 8158.37 kg/hm^2. 8719.55 kg/hm^2, and 9084.42 kg/hm^2, respectively. Compared with manual fertilization, the traditional empirical method increased yield by 6.78%, while the variable-controlled technique achieved a 4.18% yield increase. All treatment groups applied approximately 15% less fertilizer per unit area than the conventional balanced fertilization regimen. In 2018, the average yields of the manual fertilization, traditional empirical mechanical sowing, and variable-controlled mechanical sowing areas were 7995.57 kg/hm^2, 8209.72 kg/hm^2, and 8880.65 kg/hm^2, respectively. The traditional empirical method showed a 2.68% yield advantage over manual fertilization, whereas the

Table 8.3 Results of rice crop yield in test area

Year	Tests/Indicators	Artificial fertilizer spreading (kg)	Traditional empirical spreading (kg)	Real-time variable-rate spreading (kg)
2017	1	78.64	76.35	95.48
	2	86.43	94.64	86.38
	3	68.41	82.26	98.82
	4	91.18	98.45	84.67
	5	72.39	79.61	92.05
	6	92.43	91.84	87.64
	Coefficient of variation (%)	12.25	10.29	6.13
	Average yield (kg)	81.58	87.19	90.84
	Yield per hectare (kg/hm^2)	8158.37	8719.55	9084.42
	Relative yield increase (%)	6.87		4.18
2018	Average yield (kg)	79.95	82.09	88.81
	Yield per hectare (kg/hm^2)	7995.57	8209.72	8880.65
	Relative yield increase (%)	2.68		8.17
2019	Average yield (kg)	79.52	85.49	90.25
	Yield per hectare (kg/hm^2)	7952.37	8549.41	9025.03
	Relative yield increase (%)	7.51		5.56
Average yield per hectare (kg/hm^2)		8035.44	8492.89	8996.70
Relative yield increase (%)		5.69		5.93

variable-controlled technique achieved an 8.17% yield increase. In 2019, the average yield of the variable-controlled mechanical sowing area was 7952.37 kg/mu (approximately 0.067 hectares), representing a 7.51% increase compared with the manual fertilization area and a 5.56% increase compared with both the traditional empirical mechanical sowing and balanced mechanical sowing areas. Annual yield fluctuations across different experimental plots remained relatively stable, indicating consistent performance of the fertilization models.

Analysis of the 2017–2019 yield data for various regions showed that the three-year average rice yields of the manual fertilization area, traditional empirical mechanical broadcasting area, and variable-rate mechanical broadcasting area were 8035.44 kg/hm^2, 8492.89 kg/hm^2, and 8996.70 kg/hm^2, respectively. The average yield of the traditional empirical mechanical broadcasting area was 5.69% higher than that of the manual fertilization area, while the three-year average yield of the variable-rate mechanical broadcasting area was 5.93% higher than that of the traditional empirical balanced broadcasting area. These results confirm that the newly developed centrifugal real-time variable-rate fertilizer spreader can perform on-demand online variable-rate fertilization based on crop growth information. This avoids local excessive fertilizer accumulation, reduces chemical fertilizer pollution to crops and the environment, improves the population structure and spatial uniformity of crop growth, increases crop yield and plot yield balance, reduces fertilizer input, and enhances fertilizer utilization efficiency. Thus, compared with other tested fertilization methods, the newly developed fertilizer spreader exhibits superior economic, ecological, and social benefits.

8.3.4 Analysis of Partial Economic Benefit

Based on the aforementioned efficiency analysis of real-time variable-rate fertilization (VRF) operations for rice production, VRF reduced the application rate of top-dressing tillering fertilizer and production costs, balanced soil nutrient structure, decreased fertilization operation frequency, lowered labor costs, and improved fertilization efficiency. It also optimized the crop growth population structure, facilitated the formation of a high-yield crop framework, increased crop yields, and boosted planting income and profits. Key economic benefit indicators are presented in Table 8.4 (calculated using three-year average values).

Table 8.4 Results of key economic indicators

Items		Artificial fertilizer spreading	Traditional empirical spreading	Real-time variable-rate spreading
Costs	Fixed cost (RMB)	A	A + 40,000	A + 100,000
	Fertilizer input (kg/hm²)	300	300	252.36
	Fertilizer price (RMB /kg)	2.00	2.00	2.00
	Fertilizer cost (RMB /hm²)	600	600	504.72
	Labor number	8	1	1
	Labor cost (RMB / day)	50 (8hm²)	200 (54hm²)	300 (54hm²)
	Total labor cost (RMB)	400	200	300
Benefits	Rice yield (kg/hm²)	8035.44	8492.89	8996.70
	Price of rice (RMB /kg)	2.00	2.00	2.00
	Planting income (RMB /hm²)	16070.88	16985.78	17993.40

In the table, *A* denotes other common input costs in rice production; the purchase cost of the traditional balanced fertilizer spreader is 40,000 RMB, and that of the real-time variable-rate fertilizer spreader is 100,000 RMB. With the emergence of new organizations such as agricultural machinery specialized cooperatives and agricultural machinery operation service teams, the economic benefits of VRF can be analyzed excluding factors like purchase costs, equipment maintenance fees, and depreciation fees. According to the economic indicator analysis in Table 8.4, the fertilizer cost for empirical balanced fertilization (including manual spreading and traditional mechanical spreading) was 600 RMB/hm^2. In contrast, the fertilizer cost for real-time variable mechanical spreading was 504.72 RMB/hm^2, representing a saving of approximately 95.28 RMB/hm^2 compared to empirical balanced fertilization. Based on the test plot area distribution, manual fertilization required 8 workers working 6 hours per day to complete spreading across 6 plots (total 8 hm^2), with a labor cost of approximately 400 RMB/day (machinery and equipment costs are not considered here, though long-term, large-scale use would yield more pronounced economic benefits). Considering the operating efficiency of fertilizer spreaders, one operator working 6 hours per day could complete spreading over 54 hm^2. The labor cost for traditional empirical balanced mechanical spreading was approximately 200 RMB, while that for real-time variable mechanical spreading was around 300 RMB—both lower than the labor cost of manual fertilization. In terms of rice yield benefits, the planting incomes of the manual fertilization area, traditional empirical balanced fertilization area, and real-time VRF area were 16,070.88 RMB/hm^2, 16,985.78 RMB/hm^2, and 17,993.40 RMB/hm^2, respectively. Compared with manual fertilization, the traditional empirical mechanical spreading area increased yield income by approximately 914.90 RMB/hm^2; the VRF area further increased yield income by about 1007.62 RMB/hm^2 compared to the traditional empirical balanced spreading area. These results confirm that real-time VRF not only reduces fertilizer input and manual operation costs and improves spreading efficiency but also increases rice yield, delivering significant economic benefits.

8.3.5 *Effect of Fertilizer Application Rate on the Spreading*

According to the test scheme in Sect. 2.4.1, the mean fertilizer particle mass in the transverse-distributed collection pan array was standardized and converted into the actual fertilizer application rate (kg/hm^2). The particle spreading test results for single-pass fertilizer application are shown in Fig. 8.15 and Table 8.5.

As illustrated in Fig. 8.15, the distribution pattern of fertilizer particles under single-pass operation is approximately symmetrical along the spreader's travel path and varies with the application rate. Specifically, at an application rate of 450 kg/ha, the particle distribution exhibits a distinct W-shape; at 225 kg/ha, it approximates an M-shape; and at 300 kg/ha, it presents a relatively uniform Gaussian distribution. Taking the spreader's travel path as the axis of symmetry, the coefficient of variation

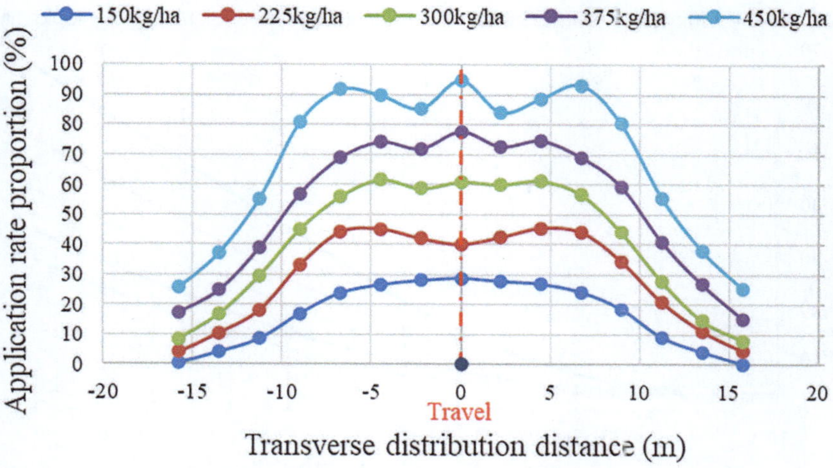

Fig. 8.15 Effect of application rate on particle distribution in single pass spreading

Table 8.5 Test results of fertilizer particle application in single pass spreading

Target fertilizer application rate/(kg/ha)	Effective working width/m	Actual fertilizer application rate /(kg/ha)		
		Mean	SD	CV/(%)
150	20.48	123.63	21.34	17.26
225	21.05	206.47	22.62	10.95
300	22.58	280.71	34.09	12.14
375	23.67	348.15	35.20	10.11
450	26.31	409.37	69.22	16.91

(CV) of particle application gradually increases as the transverse distribution distance extends outward on both sides, as shown in Fig. 8.16.

Combining the insights from Figs. 8.15 and 8.16, it is evident that when the particle distribution coefficient of variation (CV) requirement is satisfied (i.e., CV < 15%), the corresponding effective working width of the fertilizer spreader varies with different application rates—and the effective working width increases gradually as the application rate rises. Table 8.5 presents the results of the single-pass fertilizer spreading tests. At a target application rate of 150 kg/ha, the effective spreading width was 20.48 m, with an average application rate of 123.63 kg/ha within this width, a deviation of 21.34 kg/ha, and a particle distribution CV of 17.26%. At a target rate of 450 kg/ha, the effective spreading width reached 26.31 m, the average application rate within the effective width was 409.37 kg/ha, the standard deviation was 69.22 kg/ha, and the particle distribution CV was 16.91%. Notably, the CVs within the effective working width for target rates of 150 kg/ha and 450 kg/ha both exceeded the required 15%. In contrast, at target application rates of 225 kg/ha, 300 kg/ha, and 375 kg/ha, the particle distribution CVs within

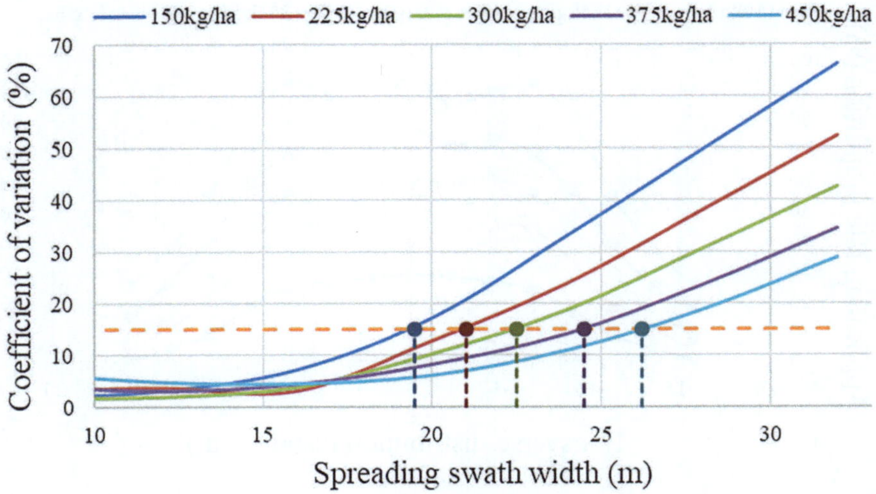

Fig. 8.16 CV of particle distribution (uniformity) within the spreading width with different fertilizer application rates

the respective effective spreading widths were 10.95%, 12.14%, and 10.11% (all below 15%), with corresponding effective spreading widths of 21.05 m, 22.58 m, and 23.67 m. Therefore, 225 kg/ha, 300 kg/ha, and 375 kg/ha are commonly recommended as the optimal target application rates for achieving superior spreading uniformity with the variable-rate fertilization (VRF) spreader.

Multi-pass fertilizer spreading performance tests were conducted at application rates of 225, 300, and 375 kg/ha. The interval between the spreader's working travel paths, corresponding to its effective spreading width, was set in accordance with the commonly adopted back-and-forth (reciprocating) operation mode. The results of the multi-pass overlapping fertilizer particle application tests are presented in Fig. 8.17.

As illustrated in Fig. 8.17, the actual application rate of multi-pass overlapping fertilization (solid orange line) is generally slightly higher than the target/ideal application rate, which aligns with previous research findings (Fulton et al., 2001). Within the effective working width of the spreader, the application rate on the right side of the left travel path compensates for that on the left side of the middle path, while the rate on the left side of the right path compensates for that on the right side of the middle path. This mutual compensation results in a multi-pass particle distribution pattern that is superior to the single-pass pattern. As shown in Fig. 8.17a, at a target/ideal application rate of 225 kg/ha and within an effective working width of 21.05 m, the actual application rate error was 2.61%, the average rate was 231.02 kg/ha, and the particle distribution coefficient of variation (CV) was 11.51%. As shown in Fig. 8.17b, at a target/ideal rate of 300 kg/ha and effective working width of 22.58 m, the actual rate error was 5.80%, the average rate was 318.47 kg/ha, and the distribution CV was 9.25%. As shown in Fig. 8.17c, at a target/ideal rate of 375 kg/

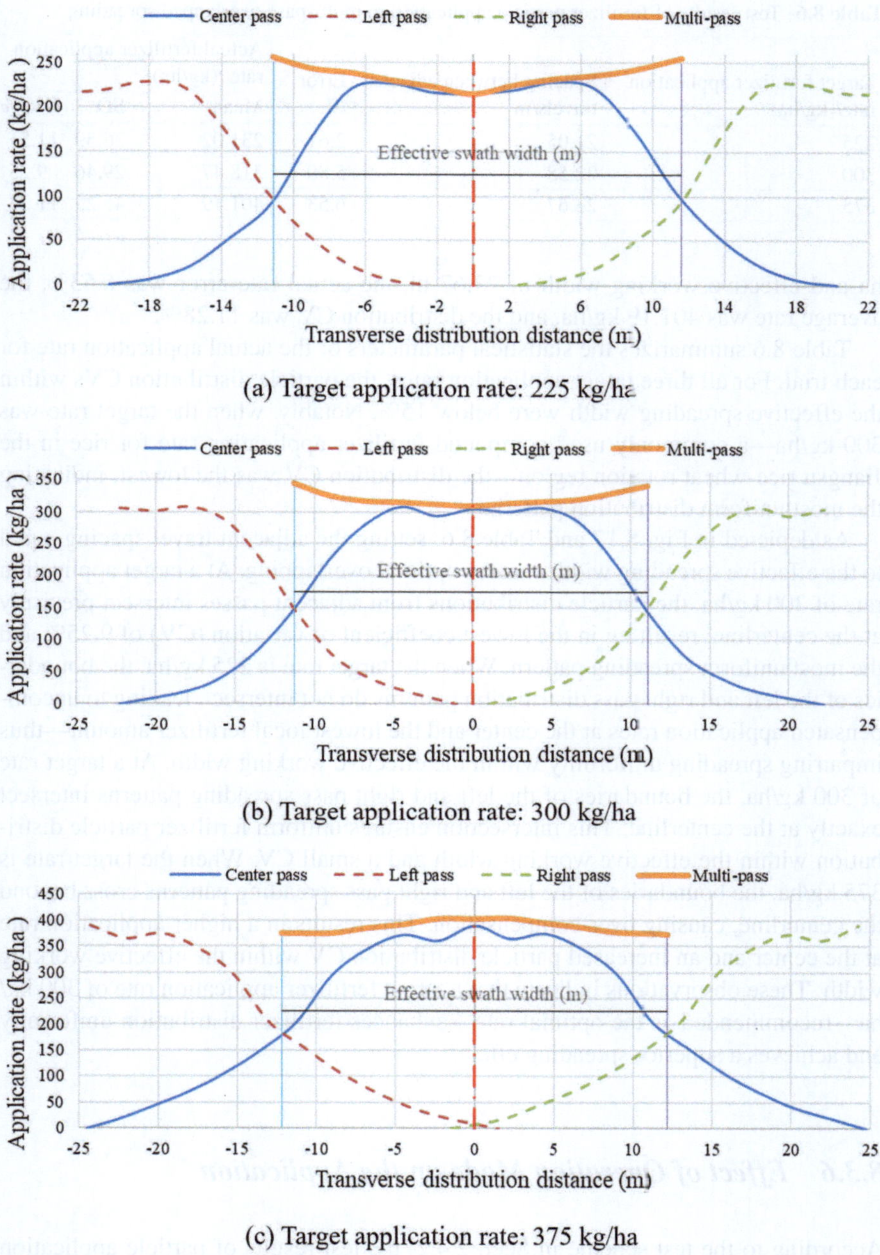

Fig. 8.17 Application of multi-pass overlapping fertilizer with different application rates. (**a**) Target application rate: 225 kg/ha. (**b**) Target application rate: 300 kg/ha. (**c**) Target application rate: 375 kg/ha

Table 8.6 Test results of fertilizer particle application in multi-pass overlapped spreading

Target fertilizer application rate/(kg/ha)	Spacing between adjacent travels/m	Error /%	Actual fertilizer application rate /(kg/ha)		
			Mean	SD	CV/%
225	21.05	2.61	231.02	26.59	11.51
300	22.58	5.80	318.47	29.46	9.25
375	23.67	6.53	401.19	45.25	11.28

ha and effective working width of 23.67 m, the actual rate error was 6.53%, the average rate was 401.19 kg/ha, and the distribution CV was 11.28%.

Table 8.6 summarizes the statistical parameters of the actual application rate for each trial. For all three target application rates, the particle distribution CVs within the effective spreading width were below 15%. Notably, when the target rate was 300 kg/ha—a commonly used compound fertilizer application rate for rice in the Jiangsu rice-wheat rotation region—the distribution CV was the lowest, indicating the most uniform distribution pattern.

As depicted in Fig. 8.17 and Table 8.6, setting the adjacent travel spacing equal to the effective spreading width ensures optimal overlapping. At a target application rate of 300 kg/ha, the particle distributions from adjacent passes intersect precisely at the centerline, resulting in the lowest coefficient of variation (CV) of 9.25% and the most uniform spreading pattern. When the target rate is 225 kg/ha, the boundaries of the left and right pass distribution patterns do not intersect, leading to uncompensated application rates at the center and the lowest local fertilizer amount—thus impairing spreading uniformity within the effective working width. At a target rate of 300 kg/ha, the boundaries of the left and right pass spreading patterns intersect exactly at the centerline. This intersection ensures uniform fertilizer particle distribution within the effective working width and a small CV. When the target rate is 375 kg/ha, the boundaries of the left and right pass spreading patterns cross beyond the centerline, causing over-compensation. This results in a higher application rate at the center and an increased particle distribution CV within the effective working width. These observations indicate that a target fertilizer application rate of 300 kg/ha—recommended as the optimal rate—enhances fertilizer distribution uniformity and achieves a superior spreading effect.

8.3.6 Effect of Operation Mode on the Application

According to the test scheme in Sect. 2.4.2, the test results of particle application under different operating mode conditions were obtained as shown in Fig. 8.18.

No significant difference in particle distribution uniformity was observed for multi-pass overlapping fertilization across different adjacent pass operating modes. Figure 8.18a illustrates the back-and-forth mode (i.e., adjacent passes in opposite directions, with blue, orange, and purple solid arrows denoting Travel N − 1, Travel

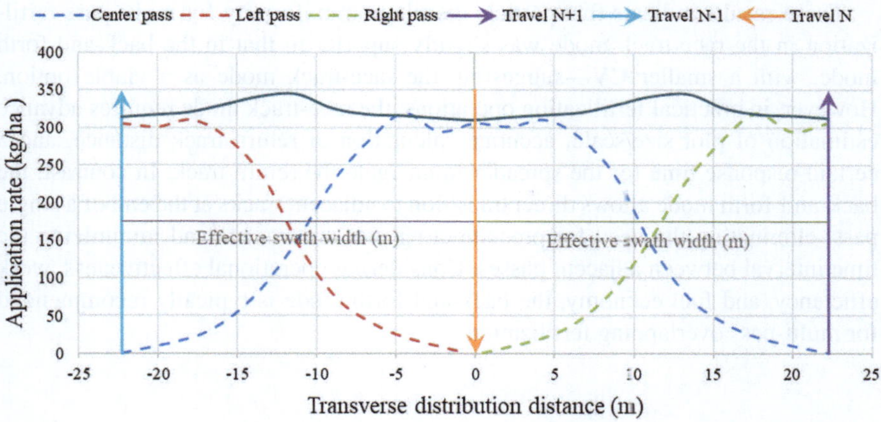

(a) Back and forth pattern (adjacent passes in reverse direction)

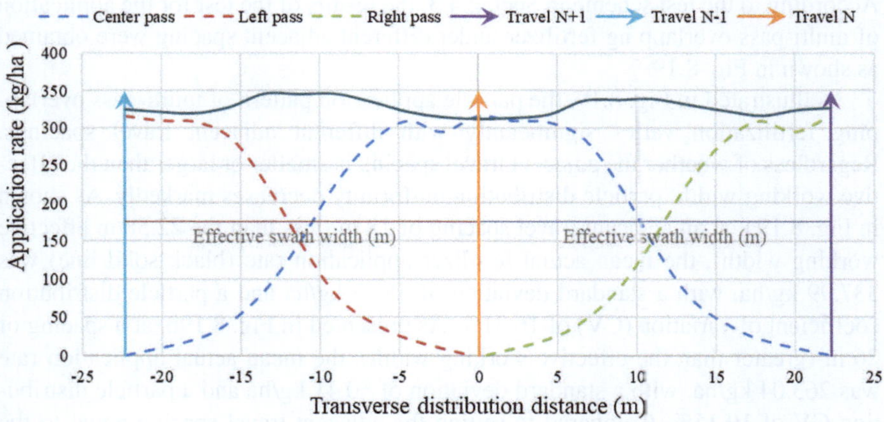

(b) Race track pattern (adjacent passes in the same direction)

Fig. 8.18 Application of multi-pass overlapping fertilizer under different travel models. (**a**) Back and forth pattern (adjacent passes in reverse direction). (**b**) Race track pattern (adjacent passes in the same direction)

N, and Travel N + 1 directions, respectively). Here, the mean actual fertilizer application rate (brown solid line) was 314.11 kg/ha, with a standard deviation of 27.19 kg/ha and a particle distribution coefficient of variation (CV) of 8.66%. Figure 8.18b shows the race-track mode (i.e., adjacent passes in the same direction, with blue, orange, and purple solid arrows indicating Travel N − 1, Travel N, and Travel N + 1 directions, respectively). The mean actual application rate was 315.90 kg/ha, with a standard deviation of 26.08 kg/ha and a particle distribution CV of 8.25%.

These results indicate that particle distribution uniformity for multi-pass fertilization in the race-track mode was slightly superior to that in the back-and-forth mode, with a smaller CV—suggesting the race-track mode as a viable option. However, in practical fertilization operations, the race-track mode requires advance estimation of plot size/scale, accurate calculation of return track distance, and a certain response time for the spreader to navigate the return track. In contrast, the back-and-forth mode allows direct transition to adjacent tracks at the end of a single pass, eliminating the need for precomputing plot dimensions and minimizing the time interval between adjacent passes. Considering operational effectiveness, work efficiency, and fuel economy, the back-and-forth mode is typically recommended for multi-pass overlapping fertilization.

8.3.7 Effect of Adjacent Travel Spacing on the Application

According to the test scheme in Sect. 2.4.3, the results of the test for the application of multi-pass overlapping fertilizer under different adjacent spacing were obtained as shown in Fig. 8.19.

As illustrated in Fig. 8.19, the particle application pattern of multi-pass overlapping fertilization varies significantly with different adjacent travel spacings. Regardless of whether the adjacent travel spacing is smaller or larger than the effective working width, particle distribution uniformity decreases markedly. As shown in Fig. 8.19a, at an adjacent travel spacing of 18 m (less than the 22.58 m effective working width), the mean actual fertilizer application rate (black solid line) was 337.99 kg/ha, with a standard deviation of 55.13 kg/ha and a particle distribution coefficient of variation (CV) of 16.31%. As presented in Fig. 8.19b, at a spacing of 26 m (greater than the effective working width), the mean actual application rate was 263.04 kg/ha, with a standard deviation of 50.41 kg/ha and a particle distribution CV of 19.16%. Compared to setting the adjacent travel spacing equal to the 22.58 m effective working width (Fig. 8.18a), both the actual application rate error (4.49%) and particle distribution CV (8.66%) increased significantly, accompanied by a decline in fertilizer distribution uniformity.

Further analysis indicates that when the adjacent travel spacing is smaller than the effective working width, the actual working area is reduced. With the fertilizer discharge flow rate (discharge amount) remaining constant, the per-unit-area application rate increases—resulting in an actual fertilizer application rate higher than the target rate, with an error of 11.24%. The boundary junction of spreading patterns between adjacent passes shifts inward, and the increased single-pass particle amount at the overlapping intersection leads to excessive mutual compensation, causing a surge in the actual application rate. This forms convex points in the particle spreading pattern, which are roughly symmetrical on either side of the central travel line. Additionally, on the central travel symmetry line, the boundaries of spreading patterns from left and right passes intersect, leading to repeated compensation of the actual application rate at this location and the formation of small convex points.

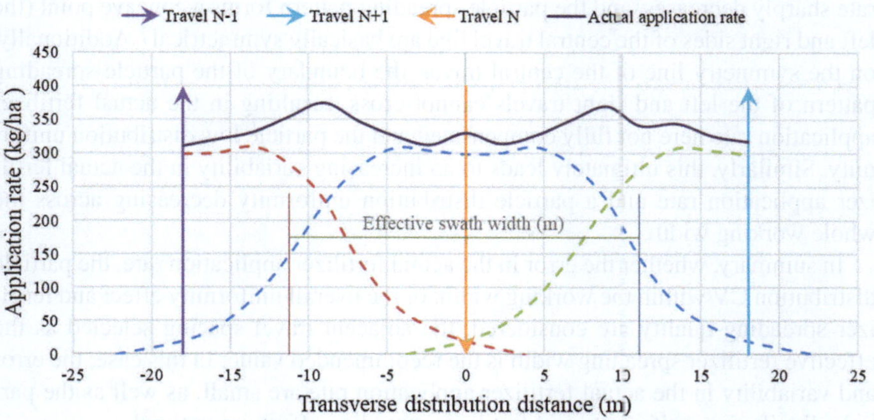

(a) Back and forth pattern (adjacent passes in reverse direction)

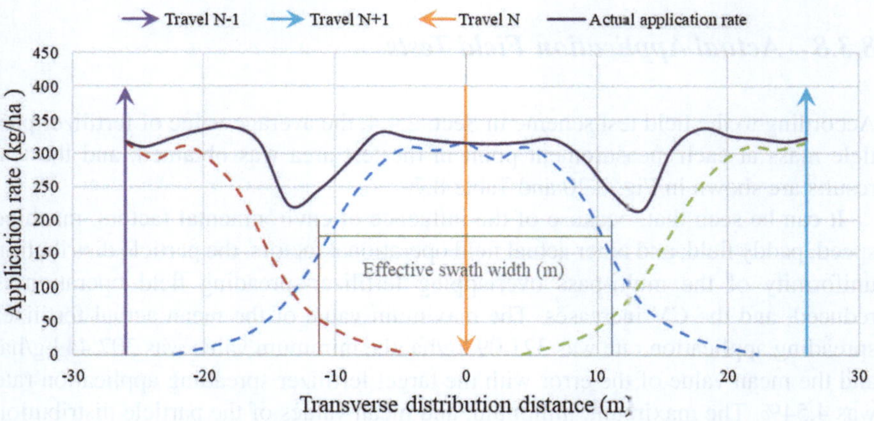

(b) Race track pattern (adjacent passes in same direction)

Fig. 8.19 Deposition distribution of multi-pass overlapping fertilizer under different adjacent spacing. (**a**) Back and forth pattern (adjacent passes in reverse direction). (**b**) Race track pattern (adjacent passes in same direction)

Ultimately, variability in the actual application rate increases, and particle distribution uniformity declines across the entire working width.

When the spacing is higher than the effective working width, the actual working area becomes larger, and the fertilizer discharge flow rate (fertilizer discharge amount) decreases the fertilizer application rate per unit area; thus, the actual fertilizer application rate is lower than the target application rate, with an error of 12.32%. The boundary junction of particle spreading patterns for adjacent travels expands outward, and the single-pass particle application amount decreases at the intersection, with lower mutual compensation. Therefore, the actual fertilizer application rate sharply decreases, and the particle spreading pattern forms a concave point (the left and right sides of the central travel line are basically symmetrical). Additionally, on the symmetry line of the central travel, the boundary of the particle spreading pattern of the left and right travels cannot cross, resulting in the actual fertilizer application rate here not fully compensated and the particle low distribution uniformity. Similarly, this ultimately leads to an increasing variability in the actual fertilizer application rate and a particle distribution uniformity decreasing across the whole working width.

In summary, whether the error in the actual fertilizer application rate, the particle distribution CV within the working width, or the overall uniformity effect and fertilizer spreading quality are considered, the adjacent travel spacing selected as the effective fertilizer spreading width is the recommended value. In this case, the error and variability in the actual fertilizer application rate are small, as well as the particle distribution uniformity and fertilizer spreading effect are optimal.

8.3.8 Actual Application Field Tests

According to the field test scheme in Sect. 2.4.4, the average value of fertilizer particle mass at each measurement point in the test area was obtained, and the test results are shown in Fig. 8.20 and Table 8.7.

It can be seen that, because of the influence of environmental factors, machine speed, paddy field, and other actual field operational factors, the particle distribution uniformity of the multi-pass overlapping fertilizer-spreading field operation is reduced, and the CV increases. The maximum value of the mean actual fertilizer spreading application rate was 321.09 kg/ha, the minimum value was 307.44 kg/ha, and the mean value of the error with the target fertilizer spreading application rate was 4.54%. The maximum, minimum, and mean values of the particle distribution CV within the working width were 13.25%, 10.95%, and 11.94%, respectively, which are all below 15%, which meets the requirements of general centrifugal fertilizer-spreading machinery for particle transverse distribution and can be used for multi-pass overlapping fertilizer spreading in actual large-scale rice cultivation in rice-wheat crop rotation regions.

8.4 Discussions

Variable-rate fertilization (VRF) technology, a core component of precision agriculture with substantial implications for agricultural production, was subjected to field trials conducted from 2017 to 2019 in rice-wheat rotation regions of the middle and lower reaches of the Yangtze River. Employing a self-developed centrifugal real-time variable-fertilizer device, the study monitored adjustments to fertilization parameters, dynamics of crop growth, and variations in yield. These results offer

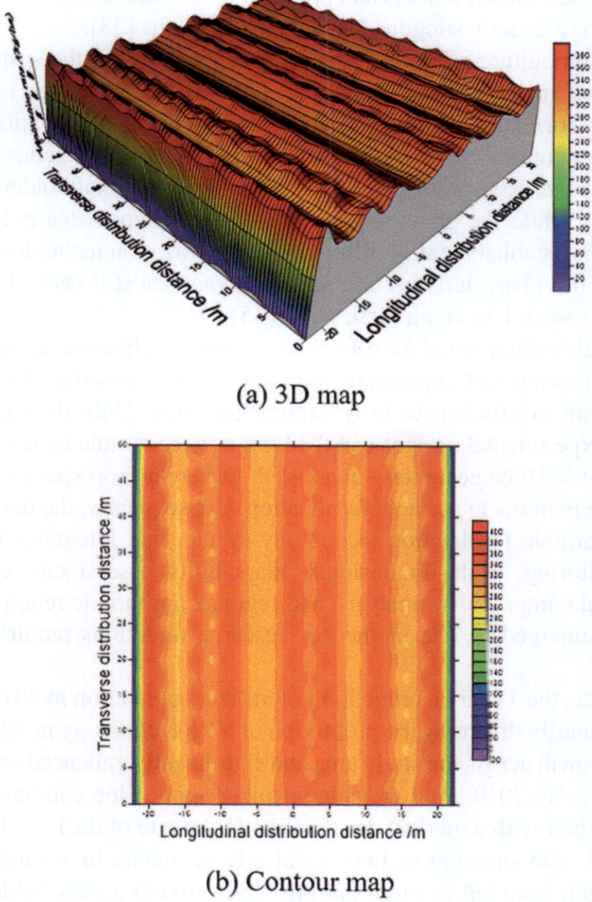

Fig. 8.20 3D and contour maps of field operation tests of multi-pass overlapping fertilizer. (**a**) 3D map. (**b**) Contour map

Table 8.7 Field test results of fertilizer particle deposition distribution in multi-pass overlapped spread

Test	Target fertilizer application rate/(kg/ha)	Spacing between adjacent travels/m	Error /%	Actual fertilizer application rate /(kg/ha)		
				Mean	SD	CV/%
1	300	22.58	2.42	307.44	39.53	12.86
2			5.64	317.92	36.27	11.41
3			2.99	309.26	40.97	13.25
4			5.07	316.03	35.54	11.24
5			6.56	321.09	35.16	10.95
Mean	/		4.54	/		11.94

critical theoretical underpinnings and practical guidelines for the implementation of VRF technology under analogous ecological conditions [33].

Precision agriculture constitutes the primary pathway for the sustainable development of modern agriculture. In intensive crop production systems, there is a distinct positive correlation between economic benefits and operational scale [9]. Traditional agricultural production in the rice-wheat rotation zones of the middle and lower Yangtze River remains largely dominated by smallholder decentralized farming [34]. While the government has rolled out supportive policies to lower input costs, the establishment of differentiated fertilization technology systems tailored to land use characteristics and nutrient dynamics still necessitates extensive experimental research to attain these goals [35].

A systematic evaluation of the effects of variable fertilization demands long-term dynamic monitoring and continuous tracking studies. Existing short-term experimental data are insufficient to fully capture its value. Only through multi-stage, progressive experimental designs can the long-term economic benefits and ecological impacts of VRF be accurately quantified. Different crop species display divergent response patterns to variable fertilization. Consequently, the development of a centrifugal variable fertilization technology system that integrates real-time crop growth monitoring, multi-dimensional adaptability assessment, comprehensive environmental compatibility analysis, and regional agronomic requirement characteristics has emerged as one of the key research directions requiring immediate attention.

For instance, the 11.85 kg reduction in fertilizer application in 2018 compared to 2017 was primarily driven by the promotion of VRF technology in 2017, which balanced crop growth across the study area and significantly enhanced overall fertilization efficiency. By 2019, total variable fertilizer application continued to decrease by 3.03 kg, albeit with a marked deceleration in the rate of decline. This trend may be associated with ongoing technological advancements in variable fertilization methods, which have led to more uniform crop growth across fields and reduced variability in fertilizer demand. Despite these changes, annual data comparisons reveal that the decline in fertilization rate has not reached statistical significance. Future research should further investigate the impact magnitude and long-term trends of this technology on fertilization rates if its application is sustained. Additionally, the potential for fertilizer application amounts to first fluctuate reversibly and then gradually stabilize during specific crop growth stages warrants in-depth exploration based on significance test results.

8.5 Conclusions

This study developed a centrifugal variable-rate fertilization (VRF) device with independent intellectual property rights, leveraging dynamic monitoring data of rice and wheat growth cycles. Based on three consecutive years of field trials, the device's performance in nitrogen application rate, crop growth, and yield was evaluated through comparisons with traditional fertilization methods. The research comprehensively assessed the operational efficiency of this technology, aiming to identify VRF techniques suitable for precision agriculture and explore their economic value and promotion potential in the rice-wheat rotation regions of the middle and lower reaches of the Yangtze River.

Field trial data revealed that the fertilizer application rate of the centrifugal VRF device decreased annually, with a per-mu reduction of approximately 15.88% compared to traditional empirical balanced fertilization. After tillering-stage fertilization, rice plants in the mechanized fertilization area exhibited minimal inter-plant growth variation and more uniform population distribution. The coefficient of variation (CV) of the normalized differential vegetation index (NDVI), derived from rice spectral data, fluctuated between 7.00% and 9.00%. Relative to the control areas, the VRF zone achieved a 5.93% increase in rice yield and saved up to 95.28 yuan per mu in costs per operation. Comprehensive analysis confirmed significant economic advantages, including an hourly yield-related income increase of 1007.62 yuan.

Empirical studies verify that real-time variable-frequency fertilization technology has distinct advantages in optimizing crop growth conditions and improving plant community structure, providing reliable support for yield enhancement. The implementation of precision control strategies has significantly reduced chemical fertilizer consumption while effectively lowering operational costs and environmental pollution risks. This technology achieves marked improvements in fertilizer use efficiency, enabling resource optimization and enhancing both product quality and market stability. Compared with traditional fertilization methods, this energy-efficient and environmentally friendly approach demonstrates superior performance, with clear advantages in economic, ecological, and social benefits. It serves as a critical reference for standardizing the promotion of variable-frequency fertilization technology in precision agriculture.

This study confirms the potential of the self-developed dual-disc centrifugal VRF spreader to improve fertilizer distribution uniformity and accuracy in precision agriculture. Optimal performance was observed at a target application rate of 300 kg/ha, with an effective spreading width of 22.58 m and a CV of 9.25% during multi-pass operations. The back-and-forth operation mode was recommended for its balanced efficiency and practicality. Field tests validated the spreader's real-world applicability, with an error rate of 4.54% and a CV of 11.94%, both meeting industry standards.

These findings contribute to the advancement of precision agriculture by providing actionable insights into multi-pass overlapping operations and developing equipment capable of reducing environmental impact, minimizing fertilizer waste, and boosting yields. Future research should focus on evaluating the spreader's performance under diverse environmental conditions, testing its adaptability to different crop systems, and integrating advanced control algorithms for further optimization.

References

1. Bierman PM, Rosen CJ, Venterea RT et al (2012) Survey of nitrogen fertilizer use on corn in Minnesota. Agric Syst 109(6):43–52
2. Bouroubi Y, Tremblay N, Vigneault P et al (2011) Fuzzy logic approach for spatially variable nitrogen fertilization of corn based on soil, crop and precipitation information [C]. In: Computational science and its application, pp 356–368
3. Chen M, Shi YY, Wang XC et al (2015) Design and experiment of variable rate fertilizer applicator based on crop canopy spectral reflectance. Trans Chin Soc Agric Mach 46(5):26–32
4. Chen SF, Zhang SP, Sun XZ et al (2012) Design and experiment of self-propelled high-ground-clearance spreader for paddy variable-rate fertilization. Trans Chin Soc Agric Eng 28(11):16–21
5. Cunningham FM (1963) Performance characteristics of bulk spreaders for granular fertilizer. Trans ASABE 6(2):108–114
6. Dammer KH, Wollny J, Giebel A (2008) Estimation of the Leaf Area Index in cereal crops for variable rate fungicide spraying. Eur J Agron 28(3):351–360
7. Ehlert D, Schmerler J, Voelker U (2004) Variable rate nitrogen fertilisation of winter wheat based on a crop density sensor. Precis Agric 5(3):263–273
8. Guo JX, Kong YL, Xie KL et al (2016) Effects of nutrient management on yield and nitrogen use efficiency of direct seeding rice. Acta Agron Sin 42(7):1016–1025
9. Jiang AN, Huang WJ, Zhao CJ et al (2007) Effects of variable nitrogen application based on characteristics of canopy light reflectance in wheat [J]. Sci Agric Sin 40(9):1907–1913
10. Li R, Ding YQ, Yu HF et al (2017) Construction and application of the electronic prescription map system of variable rate fertilization in southern Jiangsu. J China Agric Univ 22(8):124–132
11. Liu BH, Chen XP, Cui ZL et al (2015) Research advance in yield potential and yield gap of three major cereal crops. Chin J Eco-Agric 23(5):525–534
12. Liu CL, Yuan J, Liu JZ et al (2010) ARM and DSP-based bivariable fertilizing control system design and implementation. Trans Chin Soc Agric Mach 41(S1):233–238
13. Liu H, Tang L, Zhang WY et al (2009) Construction and implementation of model-based visual rice growth system. Trans Chin Soc Agric Eng 25(9):148–154
14. Maleki MR, Mouazen AM, Ketelaere BD et al (2008) On-the-go variable-rate phosphorus fertilization based on a visible and near-infrared soil sensor. Biosyst Eng 99(1):35–46
15. Maleki MR, Ramon H, Baerdemaeker JD et al (2008) A study on the time response of a soil sensor-based variable rate granular fertiliser applicator [J]. Biosyst Eng 100(2):160–166
16. Meng ZJ, Zhao CJ, Fu WQ et al (2011) Prescription map identification and position lag calibration method for variable rate application of fertilizer [J]. Trans Chin Soc Agric Mach 42(7):204–209
17. Ni J, Wang TT, Yao X et al (2013) Design and experiments of multi-spectral sensor for rice and wheat growth information. Trans Chin Soc Agric Mach 44(5):207–212
18. Patterson DE, Reece AR (1962) The theory of the centrifugal distributor. I: motion on the disc, near-centre feed. J Agric Eng Res 7(3):232–240
19. Radite PAS, Umeda M, Iida M, et al (2000) Variable rate fertilizer applicator for paddy field [C]. Milwaukee, Wisconsin: 1–11
20. Rotundo José L, Rech R, Cardoso MM et al (2022) Development of a decision-making application for optimum soybean and maize fertilization strategies in Mato Grosso. Comput Electron Agric 193:106659
21. Shi YY, Chen M, Wang XC et al (2018) Design and experiment of variable-rate fertilizer spreader with centrifugal distribution cover for rice paddy surface fertilization. Trans Chin Soc Agric Mach 49(3):86–93
22. Shi YY, Chen M, Wang XC et al (2018) Numerical simulation of spreading performance and distribution pattern of centrifugal variable-rate fertilizer applicator based on DEM software. Comput Electron Agric 144:249–259

23. Shi YY, Hu ZC, Wang XC et al (2018) Fertilization strategy and application model using a centrifugal variable-rate fertilizer spreader [J]. Int J Agric Biolog Eng 11(6):41–48
24. Shi YY, Zhu Y, Wang XC et al (2020) Progress and development on biological information of crop phenotype research applied to real-time variable-rate fertilization. Plant Methods 16:11
25. Song XY, Wang JH, Huang WJ et al (2009) Monitoring spatial variance of winter wheat growth and grain quality under variable-rate fertilization conditions by remote sensing data. Trans Chin Soc Agric Eng 25(9):155–162
26. Sun GX, Chen M, Wang XC et al (2014) Research status of key techniques of variable rate fertilizer applicator. Jiangsu Agric Sci 42(2):333–339
27. Tang H, Wang JW, Xu CS et al (2019) Research progress analysis on key technology of chemical fertilizer reduction and efficiency increase [J]. Trans Chin Soc Agric Mach 50(4):1–19
28. Tian YH, Yin B, Zhu ZL (2006) Study on nitrogen losses via leaching in paddy soil. J Anhui Agric Sci 34(2):2792–2794
29. Wang J, Wang DJ, Zhang G (2010) Effect of incorporated straw-nitrogen fertilizer on nutrient leaching in paddy soils. Chin J Eco-Agric 18(2):316–321
30. Wang JW, Pan ZW, Zhou WQ et al (2015) Design and test of SYJ-2 type liquid variable fertilizer. Trans Chin Soc Agric Mach 46(7):53–58
31. Yu F, Shi WM (2015) Nitrogen use efficiencies of major grain crops in China in recent 10 years. Acta Pedol Sin 52(6):1311–1324
32. Yuan J, Liu CL, Li YM et al (2010) Gaussian processes based bivariate control parameters optimization of variable-rate granular fertilizer applicator. Comput Electron Agric 70(1):33–41
33. Zaman QU, Esau TJ, Schumann AW et al (2011) Development of prototype automated variable rate sprayer for real-time spot-application of agrochemicals in wild blueberry fields[J]. Comput Electron Agric 76(2):175–182
34. Zhang R, Wang X, Zhao CJ et al (2012) Design and experiment of variable rate fertilizer spreader with conveyor chain. Trans Chin Soc Agric Eng 28(6):20–25
35. Zheng WW, He ZZ, Xu GY (2017) Evaluation of over fertilization in main grain crops in Jiangsu and its influencing factors. Res Agric Modern 38(4):666–672

MIX
Papier aus verantwortungsvollen Quellen
Paper from responsible sources
FSC® C105338

If you have any concerns about our products,
you can contact us on
ProductSafety@springernature.com

In case Publisher is established outside the EU,
the EU authorized representative is:
Springer Nature Customer Service Center GmbH
Europaplatz 3, 69115 Heidelberg, Germany

Printed by Libri Plureos GmbH
in Hamburg, Germany